Teacher's Resource

for use with

Technology

by

R. Thomas Wright
Professor Emeritus, Industry and Technology
Ball State University
Muncie, Indiana

Publisher
The Goodheart-Willcox Company, Inc.
Tinley Park, Illinois
www.g-w.com

Manufactured in the United States of America.

Teacher's Manual
ISBN-13: 978-1-59070-721-0
ISBN-10: 1-59070-721-4

Teacher's Resource Binder
ISBN-13: 978-1-59070-722-7
ISBN-10: 1-59070-722-2

Teacher's Resource CD
ISBN-13: 978-1-59070-723-4
ISBN-10: 1-59070-723-0

1 2 3 4 5 6 7 8 9 – 08 – 11 10 09 08 07 06

Contents

Introduction

Technology and Technology Education—A Background

People live in a human-built world. We eat food that was raised and processed using technology. We live in constructed homes and work in constructed offices, stores, and factories. We use energy through thousands of technological devices. We use thousands of manufactured products, which range from the simple wooden pencil to complex computer systems. We communicate using technological devices, systems, and products such as printed and electronic media. We maintain our health and treat illnesses using technology. We move freely on transportation vehicles, and we receive goods through transportation systems. Our life is indeed enriched because of technology.

Not all the results of technology are positive, however. People using technology have polluted our water and air. Technological devices such as wind generators and cellular telephone towers mar the beauty of the landscape. Communication media shape our attitudes. Neighborhoods are impacted by transportation systems. Construction projects can change some people's entire way of life.

Without an adequate knowledge of technology, people are afraid of the results of technological change. The hope for the future lies in a populace that can understand and direct the path of technological development. A relatively new school discipline, technology education, is accepting the challenge of teaching this understanding. Technology education interprets the practices of designing, operating, and using technology and the impacts of these actions on people, society, and the environment.

Through technology education experiences, students discover that the word *technology* has different meanings in different contexts and that different people see it differently. Specifically, students learn that technology can be described as products (computers, lasers, fiber-optic communication systems), as processes (computer-aided design [CAD], nuclear power generation, robotic assembly), or as organizations (industrial organizations, company structures), Figure IN-1.

Furthermore, through the study of technology, students are able to differentiate between technology and other areas of human knowledge and practice, such as science. Students discover that science searches for laws and principles that explain the natural world. Technology, in contrast, explores how people develop and use tools, materials, and systems to control and live in harmony with natural and human-made environments, Figure IN-2.

Technology education presents technology as a systematic human activity. It shows that all technologies have inputs, processes, and outputs. Technology education demonstrates

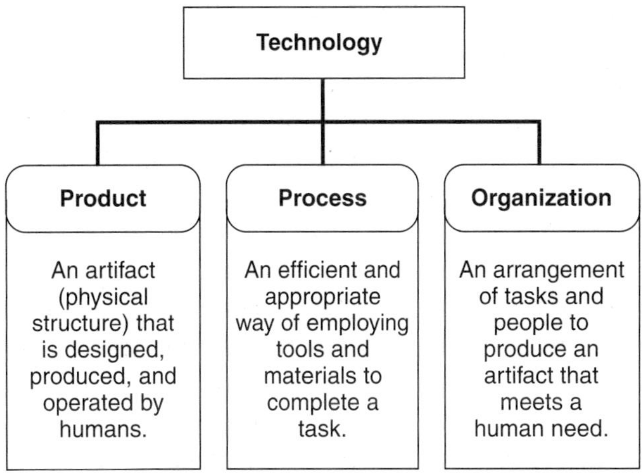

Figure IN-1

Technology . . .

... is human knowledge about using tools, materials, and systems to produce artifacts (outputs) that modify or control the environment.

Figure IN-2

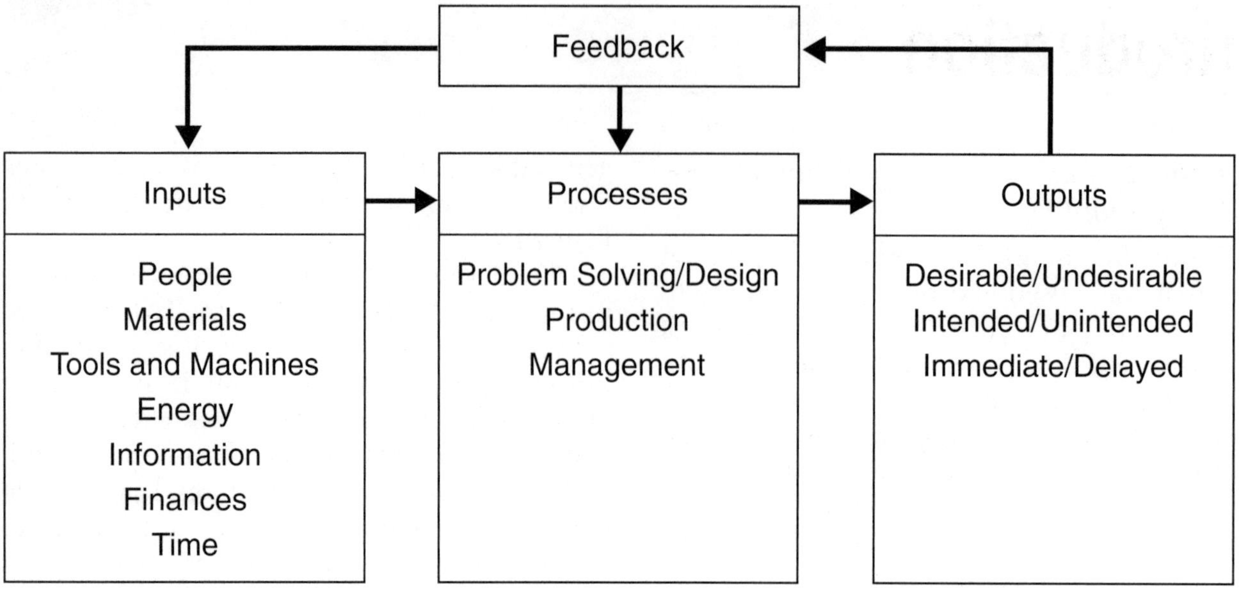

Figure IN-3

how technological systems use feedback to ensure that a system operates within established parameters, Figure IN-3. Finally, technology is presented as purposeful human action. Technology is the result of human volition. Technology is developed to meet an identified human want or need in mind.

Technology can be viewed in its entirety. Questions such as "What is technology?" "How does technology impact human life?" and "What is a technological system?" can be explored. This is useful information. This broad approach, however, does not allow students to explore specific technologies. For this reason, most technology educators will first introduce a broad view of technology and then explore several contexts in which technology is applied. These contexts allow students to explore a number of technological systems. One common grouping of technologies views them as human productive activities. This grouping explores those technological actions that people have done, are doing, and most likely will be doing in the future. These groupings are agriculture, construction, information and communication, energy conversion, manufacturing, medical, and transportation technologies, Figure IN-4. This approach proposes that people have used, are using, and will use technology to grow and process foods, communicate and process information, construct structures, convert and apply energy, promote wellness, treat illnesses, manufacture products, and transport cargo.

Technology Systems—An Instructional Approach

This instructional package has been developed to allow students systematically to study technology systems as phenomena that directly impact their lives and to study technology as something they have the power to develop, use, and control. The package includes three important teaching resources:

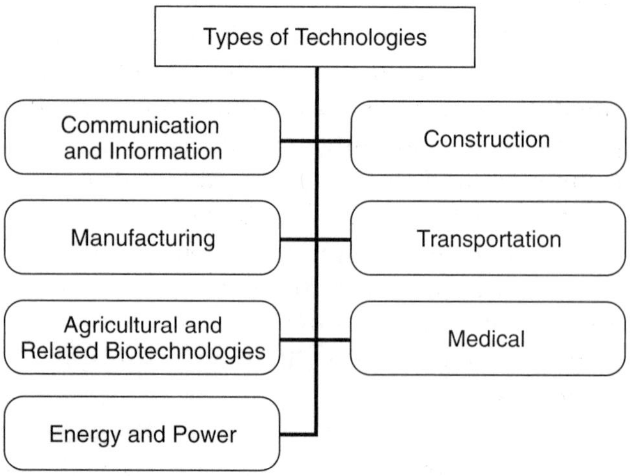

Figure IN-4

- *Technology,* which includes information about technology and its various contexts, lists of key words, review questions, and laboratory activities.
- The teacher's resource, that is, the *Teacher's Manual,* the *Teacher's Resource Binder,* or the *Teacher's Resource CD.* Each provides background information for each teaching section, suggestions for implementing student activities, review quizzes, reproducible masters, and other materials to help teachers present their course. In addition, the binder and the CD include transparency masters.
- *Student Activity Manual,* which includes forms that support the activities in the textbook.

These materials have been organized into teaching sections that correspond to the sections of the student textbook. Each section covers several chapters in the textbook. Each section has a specific theme and includes an estimate of the time needed to cover the material. This structure is shown in Figure IN-5.

36-Week (Two-semester) Schedule					
	Section	Focus	Text Chapters	Activities	Weeks
Semester I	1	Technology	1–3	1A & B	1
	2	Systems	4–6	2A & B	1
	3	Tools	7–8	3A & B	3
	4	Design	9–12	4A & B	5
	5	Production	13–18	5A–D	6
		Unassigned time			2
Semester II	6	Communication	19–23	6A & B	4
	7	Transportation	24–26	7A & B	3
	8	Energy	27–28	8A & B	3
	9	Agriculture and Medicine	29–31	9A & B	3
	10	Management	32–34	10A & B	3
	11	Society	35–36	11A & B	1
		Unassigned time			1

Figure IN-5

The instructional system for the student of technology has been developed with specific roles for the teacher, students, textbook, and activities in mind. These roles are summarized in the following.

Role of the Student

School programs are designed to help students prepare for meaningful and rewarding participation in society. The roles students play in their various classes during a typical school day are markedly different. Technology education has some features that make it a unique subject, Figure IN-6. One of these features relates to the role of students in approaching the course content. Students are considered active participants. Students are not viewed as passive receptors of wisdom. Students in technology education are included in the decision-making process. The teacher sets the boundaries of the study, but the students are involved in developing the structure and the activities that exist within these boundaries. For example, students do not have a choice on whether to study construction technology. They should be directly involved, however, in deciding which structures they will model during the study. This approach requires students to be accountable for their learning and directed in their actions.

Role of the Teacher

Teachers of technology education must have three areas of competence. First, they must possess a broad understanding of technology and industry and the impacts they have on society, individuals, and the environment. Second, the teachers must possess a level of technical skill in the designing and production of products and structures, the use of tools and machines, and the maintenance of technological devices and systems. Third, technology teachers must be capable of selecting and using a wide range of teaching methods and strategies to facilitate student learning.

Technology teachers also must have a special attitude about the subject matter and students. They must believe that students can, and are willing to, learn. This belief must be communicated through an excitement about the subject

Characteristic	Description
Mission	Technology education is general education in the same way science, mathematics, and language arts are.
Focus	Technology education focuses on technological systems and their social/cultural impacts. Technical means are used to help students develop deeper understandings of the essential concepts of technology.
Emphasis	Technology education is an *activity-centered* study that helps students understand the technological portion of society by emphasizing the *why* behind the *how*.
Goal	Technology education strives to enable students to become technologically literate (able to find, interpret, apply, and evaluate information about technology).
Instructional Practices	Technology education is designed to encourage students to become problem solvers, flexible thinkers, and group participants through the use of a combination of individual and group learning activities that students help plan and carry out. Teacher-developed procedures have limited use in this program.
Thinking Mode	Technology education encourages divergent thinking (What are the WAYS we could do this task?), followed by convergent thinking (Which of their ways will work best at what point in time?).
Teacher Role	The technology teacher is a facilitator of learning, as opposed to the person who knows THE answer. He or she manages the learning environment.
Student Role	The students in technology education are encouraged to be creative and active participants. This requires teachers to have faith in people—a *can do* philosophy. Technology education has high expectations for all students.
Safety	Technology education places strong emphasis on safe laboratory practices. It views safety as both an attitude and a practice developed through teacher-student activities. Therefore, the teacher is expected to work with the students to develop essential safety rules and procedures.

Figure IN-6

matter—"Technology is fun and interesting to learn about!"—and a concern for students—"I am truly interested in you and your educational achievements!" Finally, technology teachers must realize that there are not many "right" answers. There are, usually, a group of possible answers from which an appropriate answer may be selected.

The attributes and attitudes are reflected in the classroom where the technology teacher:

- Structures the overall class and interacts with students to structure specific learning activities.
- Guides students toward appropriate solutions.

- Presents short, exciting content presentations that include student interactions.
- Uses student input to select products, structures, and systems to be fabricated and tested.
- Develops and demonstrates, with student interaction, appropriate and safe production procedures that are used to fabricate products, structures, and systems. (**Student safety** is always important and is the responsibility of the teacher. No textbook or set of rules can teach safety. They can only support teacher actions. The proper, safe operation of tools and machines should be presented and reinforced in the laboratory setting.)
- Serves as a technical adviser for students. It is better to react to student solutions than to give them the one "right" way to do a task. This approach requires the student to ask, "Do you think the design our group developed can be produced?" rather than "What are we going to make?" This approach may cause the class to cover less content, but it helps students learn the content more thoroughly. The goal of technology education is to develop key concepts about technology, not to cover large volumes of specific information that may become obsolete before it is useful to the student. This goal allows a course to sacrifice a certain amount of depth in order to cover broad content in a more meaningful way.
- Develops appropriate evaluation instruments and procedures. The teacher's resource and the textbook contain review questions and sample quizzes. Do not hesitate to develop your own instruments that more closely match your instruction methods and your students.
- Adjusts content for individual student abilities and learning styles. The problem-solving and open-ended nature of technology education activities makes this task much easier. Each instructional section of this manual provides hints for adjusting the activities to accommodate varying student abilities.

People learn at differing rates, using different approaches. This is coupled with people possessing varying degrees of eye-hand coordination (physical skill) and ability to see spatial relationships (visualization skill). These factors present the technology educator with a unique challenge—structuring the course so that every student has the opportunity to succeed.

This requires that various learning strategies be selected to reach course objectives and that these strategies be selected with student abilities in mind. The chart in Figure IN-7 presents some approaches and types of activities that can be used to meet instructional objectives. These activities should be modified to meet the needs of the gifted, the average, and students with special needs.

Helping Students Gain Basic Information	
Textbook Reading Filmstrips Slide Series Videotapes	Films Field Trips Guest Speakers
Helping Students Develop Creativity	
Product Design Structure Design	Brainstorming Advertising Design
Helping Students Develop Problem-Solving Skills	
Product Design Process Planning	Financial Planning Structure Design
Helping Students Apply Technology	
Product Design Structure Product Production System Testing	Material Processing Managing Structure Fabrication
Helping Students Develop Communication Skills	
Class Discussions Engineering Reports Laboratory Summaries	Market Surveys Product Presentations Product Sales
Helping Students Develop Leadership Skills	
Decision Making Cooperating	Managing

Figure IN-7

Gifted students present a challenge because they may complete assigned work quickly or become bored with routine assignments. These students often need to be challenged with open-ended activities that focus on societal issues and require the use of problem-solving skills and creativity.

Special needs students may require more structure, short-range goals, and activities that build upon their strengths. Class activities in technology education often involve group work, allowing these students to complete simpler, yet essential, tasks that contribute to group success and develop self-esteem. At no time should special needs students be relegated to insignificant or nonessential tasks. The laboratories have many tasks that special needs students can complete that are still important for the success of the class activity.

Each section of the teacher's resource provides suggestions for adjusting course content to accommodate the gifted and the special needs student.

Role of the Textbook

The textbook is a vital part of an instructional system. It is designed to give students the background to understand the course content. Technology education does not accept skill development and mastery of technical knowledge as a central mission. Therefore, *Technology* does not contain specific process or machine operation procedures, technical data (such as screw sizes or abrasive grits), or specific sets of safety rules to be memorized.

Technology should be used in delivering the basic content for the course and as a reference in completing laboratory activities. Students should be given reading assignments for their study periods or as homework. They should bring the book to class each day, so they can use it as a source of specific information for completing laboratory assignments.

Role of the Teacher's Resource

The teacher's resource, whether the manual, binder, or CD, is, as the name states, a resource to aid the teacher. The teacher's resource is not the final answer for presenting a course about technology systems. The final answer resides in the teacher. The teacher is ultimately responsible for selecting, delivering, and evaluating student learning activities.

To assist the teacher, this teacher's resource serves several purposes. This resource:

- Offers a rationale for teaching technology systems and for structuring the program elements.
- Presents approaches and procedures for selecting, organizing, delivering, and evaluating student learning activities.
- Lists references for teacher information and teacher development.
- Presents possible laboratory activities for the various sections in the course.
- Indicates the conceptual structure for the various topics to be covered.
- Suggests ways to accommodate students with varying abilities and interests.
- Provides reproducible masters and transparency masters (transparency masters in the binder and CD only) to aid in student interest and understanding. These materials can be used to introduce fundamental concepts, to generate discussion about the topic, and to facilitate your presentation to the class. Many of the transparencies can be reproduced as a handout as well as used as a transparency.

This teacher's resource should be read and studied well in advance of implementing a course in technology systems. At no time does this resource attempt to be a "cookbook" containing set directions, laboratory procedures, or lecture outlines for teaching the course. The teacher's resource contains suggestions to aid teachers in designing and presenting an exciting course.

This teacher's resource is not designed to fit the concept of an "ideal" laboratory, an "ideal" class, an "ideal" course length, or the illusive "average" student. It is designed with enough flexibility to facilitate a program in a variety of traditional and contemporary laboratories with students of varying abilities and interests.

Reading this teacher's resource from cover to cover prior to implementing the course provides you with:

- An overview of the course and how its parts are logically integrated.
- An opportunity to prepare long-range plans for:
  a. Lesson and activity planning.
  b. Supplemental resources.
  c. Support material (transparencies and slides, for example) development.
  d. Laboratory modification and improvement.

This teacher's resource contains features that include:

- Clearly defined instructional objectives.
- Instructional strategies that can be used in teaching situations for students of varying ability levels.
- Suggestions and directions to reinforce, extend, and enrich the learning of students.
- Instructional strategies and extensive activities that require the use of higher-order thinking skills.
- Suggestions for appropriate questions to guide student discovery and exploration.
- Answers to questions and solutions to problems.
- A variety of evaluative procedures for assessing student achievement and for documenting mastery of content. These procedures include written quizzes and laboratory experiences.
- Background information for teaching content areas in the textbook that have not traditionally been in a subject or course.
- Problems and activities that require the application of technology concepts to life skills.

Evaluation instruments (quizzes and rating scales) and procedures help the teacher address three important aspects of the program:

- Student progress and achievement.
- Teacher and teaching method effectiveness.
- Program adequacy.

Evaluating for the purpose of assigning student grades is only a portion of the value of the continuous assessment of student progress. Data gathered during evaluation allow the teacher to identify student learning difficulties, determine concepts that have been miscommunicated or misunderstood, and measure student progress toward course objectives.

Role of the Student Activity Manual

The *Student Activity Manual* for *Technology* is used up during the course. It contains many of the forms and activity sheets needed to conduct the laboratory activities. Most of the forms and activity sheets are designed to be flexible. They are structured to fit more than one product or process.

Individual teachers may want to supplement sheets in the *Student Activity Manual* with forms tailored to the specific laboratory activities they select for the course.

A significant feature of the *Student Activity Manual* is the professional-looking forms that can save the teacher many hours of preparation. These forms have been designed to allow an open program or approach.

Laboratory Activities

The goal of technology education is *technological literacy,* which contributes to the students' ability to participate in society. These contributions are made through citizenship (voting and community service involvement, for example), careers (selection and contribution), and consumer roles (purchasing and using, for example).

The laboratory activities are designed to make abstract concepts easier for the student to understand. The activities are *not* designed to develop tool skills or specific technical knowledge. That task is the purview of occupational education at vocational schools, career centers, community colleges, and universities.

The laboratory activities allow students to gain experience in solving problems. Students participate in group and individual activities involving the design, construction, operation, and maintenance of technological structures and systems. They also give students practice in communicating technological information.

Each section in this course has two types of laboratory activities—design problems and fabrication problems. Design problems use a *design brief* to communicate a situation, state a problem, or describe an opportunity. They then challenge the student to action. Students use their own creativity, problem-solving abilities, and management skills to address the design brief. Design problems do not have procedures developed and presented by the author. Rather, they have general guidelines, which are given to help teachers work with students in identifying processing and safety procedures.

Fabrication problems are more structured. The students are given a set of plans or a procedure to use investigating a specific technology. An effective way to present fabrication problems to the students is to have them record the laboratory procedures as they are demonstrated. This method allows the teacher to interact with the class, as opposed to simply dictating "the way" to do the task. Also, by having students complete procedure sheets while the demonstration is presented, they are encouraged to give close attention to the task being presented.

A final suggestion for laboratory activities is to use a combination of the design and the fabrication problems. This keeps the students challenged, and it keeps the laboratory experiences fresh and challenging for the teacher. Using the same type of activity over an extended period of time may cause students and parents to inappropriately view the course as stagnant.

The laboratory and laboratory activities are vital parts of a strong technology education program. This program has been designed with few laboratory restrictions. It can be conducted in both traditional material-centered (woodworking or metalworking) laboratories and general technology laboratories. This characteristic has been achieved by not specifying activities that "must" be conducted in order to complete the mission of the course. Each section presents a number of activities. Thus, the program is relatively easy to implement in any broadly equipped laboratory that supports material processing.

Taking Your Technology Knowledge Home Activities

Sections 3 through 10 have special activities called "Taking Your Technology Knowledge Home." These activities allow students to work with parents and guardians to apply what they learned in their technology education class to everyday challenges around their home and neighborhood. The activities can also be adapted to school.

Standards for Technological Literacy Correlation Chart

The International Technology Education Association (ITEA) and its Technology for All Americans Project developed *Standards for Technological Literacy: Content for the Study of Technology* to identify the essential core of technological knowledge and skills for students in grades K–12. This work defined twenty separate standards, divided into five broad categories. Within each standard, benchmark topics are defined for four different grade levels:

- Grades K–2
- Grades 3-5
- Grades 6–8
- Grades 9–12

The following chart lists the standards and the benchmark topics for grades 9–12. Adjacent to each standard and benchmark topic are the chapter or section number and page references identifying material in *Technology* relating to the item.

***Standards for Technological Literacy* Correlation Chart**	
Standard 1. Students will develop an understanding of the characteristics and scope of technology.	Chapter 1: 17–19 Chapter 3: 46–53 Chapter 4 Chapter 5: 82, 85–98 Chapter 9 Chapter 32: 640–641
Benchmark Topics:	
The nature and development of technological knowledge and processes are functions of the setting.	Chapter 3: 49–53 Chapter 5: 85–98
The rate of technological development and diffusion is increasing rapidly.	Chapter 1: 19 Chapter 4
Inventions and innovations are the results of specific, goal-directed research.	Chapter 4 Chapter 9
Most development of technologies these days is driven by the profit motive and the market.	Chapter 1: 17–18 Chapter 3: 46–48 Chapter 5: 82 Chapter 32: 640–641

Standards for Technological Literacy Correlation Chart

Standard 2. Students will develop an understanding of the core concepts of technology.	Chapter 1: 17–19 Chapter 2: 31–40 Chapter 5: 82–83, 99 Chapter 6: 109–118 Chapter 8: 168–170 Chapter 9 Chapter 10: 190–192 Chapter 11: 206–214, 218 Chapter 32: 641–644 Chapter 33: 661–663
Benchmark Topics:	
Systems thinking applies logic and creativity with appropriate compromises in complex real-life problems.	Chapter 10: 190–191 Chapter 11: 206–214
Systems, which are the building blocks of technology, are embedded within larger technological, social, and environmental systems.	Chapter 10: 190–191
The stability of a technological system is influenced by all of the components in the system, especially those in the feedback loop.	Chapter 2: 31–40
Selecting resources involves trade-offs between competing values, such as availability, cost, desirability, and waste.	Chapter 10: 192
Requirements involve the identification of the criteria and constraints of a product or system and the determination of how they affect the final design and development.	Chapter 5: 82–83 Chapter 9 Chapter 10: 192
Optimization is an ongoing process or methodology of designing or making a product and is dependent on criteria and constraints.	Chapter 11: 210–211, 218
New technologies create new processes.	Chapter 1: 19
Quality control is a planned process to ensure that a product, service, or system meets established criteria.	Chapter 6: 109–111 Chapter 8: 168–170 Chapter 33: 661–663
Management is the process of planning, organizing, and controlling work.	Chapter 2: 38 Chapter 5: 99 Chapter 32: 641–644
Complex systems have many layers of controls and feedback loops to provide information.	Chapter 2: 40 Chapter 6: 109–118

Standards for Technological Literacy Correlation Chart

Standard 3. Students will develop an understanding of the relationships among technologies and the connections between technology and other fields of study.	Chapter 1: 20–22 Chapter 2: 34 Chapter 3: 52 Chapter 4 Chapter 5: 97 Chapter 6: 108, 117 Chapter 7: 148 Chapter 8: 171 Chapter 9: 187 Chapter 10: 198 Chapter 11: 213–214 Chapter 12: 231 Chapter 13: 251 Chapter 14: 265 Chapter 15: 277 Chapter 17: 333 Chapter 18: 353 Chapter 19: 368, 370 Chapter 20: 389 Chapter 21: 405, 413 Chapter 22: 418–420, 423–424, 429 Chapter 23: 442–443, 451 Chapter 24: 471 Chapter 25: 501 Chapter 26: 515 Chapter 27: 530 Chapter 28: 558 Chapter 29: 584 Chapter 30: 607 Chapter 31: 621–627 Chapter 33: 668
Benchmark Topics:	
Technology transfer occurs when a new user applies an existing innovation developed for one purpose in a different function.	Chapter 4
Technological innovation often results when ideas, knowledge, or skills are shared within a technology, among technologies, or across other fields.	Chapter 1: 21–22 Chapter 4
Technological ideas are sometimes protected through the process of patenting.	Chapter 4: 61–63
Technological progress promotes the advancement of science and mathematics.	Chapter 2: 34 Chapter 3: 52 *(continued)*

***Standards for Technological Literacy* Correlation Chart**	
Technological progress promotes the advancement of science and mathematics. *(continued from the previous page)*	Chapter 4: 73 Chapter 6: 108, 117 Chapter 8: 171 Chapter 10: 198 Chapter 11: 213–214 Chapter 13: 251 Chapter 14: 265 Chapter 15: 277 Chapter 18: 353 Chapter 19: 368 Chapter 20: 389 Chapter 21: 405 Chapter 22: 423–424 Chapter 23: 442–443 Chapter 24: 471 Chapter 25: 501 Chapter 26: 515 Chapter 28: 558 Chapter 30: 607 Chapter 31: 621–627 Chapter 33: 668
Standard 4. Students will develop an understanding of the cultural, social, economic, and political effects of technology.	Chapter 1: 17–18, 20–21 Chapter 3: 44–46 Chapter 6: 105–109 Chapter 9: 185–187 Chapter 32: 644–649 Chapter 33: 653–655, 665–671 Chapter 35 Chapter 36: 716–719
Benchmark Topics:	
Changes caused by the use of technology can range from gradual to rapid and from subtle to obvious.	Chapter 3: 44–46 Chapter 6: 107–109 Chapter 35: 692–695 Chapter 36: 718–719
Making decisions about the use of technology involves weighing the trade-offs between the positive and negative effects.	Chapter 1: 20–21 Chapter 6: 105–107 Chapter 32: 644–645 Chapter 35 Chapter 36: 716–718
Ethical considerations are important in the development, selection, and use of technologies.	Chapter 9: 185–187

Standards for Technological Literacy Correlation Chart	
The transfer of a technology from one society to another can cause cultural, social, economic, and political changes affecting both societies to varying degrees.	Chapter 1: 17–18
Standard 5. Students will develop an understanding of the effects of technology on the environment.	Chapter 1: 20–21 Chapter 6: 111–113 Chapter 18: 352–354 Chapter 25: 496–498 Chapter 28: 543–544 Chapter 31: 627–629 Chapter 35: 691–692, 696–700 Chapter 36: 716–718
Benchmark Topics:	
Humans can devise technologies to conserve water, soil, and energy through such techniques as reusing, reducing, and recycling.	Chapter 18: 352–354
When new technologies are developed to reduce the use of resources, considerations of trade-offs are important.	Chapter 31: 627–629 Chapter 35: 697–699
With the aid of technology, various aspects of the environment can be monitored to provide information for decision-making.	Chapter 6: 111–113 Chapter 25: 496–498 Chapter 35: 696–700
The alignment of technological processes with natural processes maximizes performance and reduces negative impacts on the environment.	Chapter 1: 20–21 Chapter 18: 352–354 Chapter 28: 543–544
Humans devise technologies to reduce the negative consequences of other technologies.	Chapter 18: 352–354
Decisions regarding the implementation of technologies involve the weighing of trade-offs between predicted positive and negative effects on the environment.	Chapter 1: 20–21 Chapter 35: 699–700 Chapter 36: 716–718
Standard 6. Students will develop an understanding of the role of society in the development and use of technology.	Chapter 2 Chapter 4: 61–63, 75–76 Chapter 6: 104 Chapter 22: 429 Chapter 33: 653–654, 664–665, 671 Chapter 35: 692, 700–702 Chapter 36: 706–710
Benchmark Topics:	
Different cultures develop their own technologies to satisfy their individual and shared needs, wants, and values.	Chapter 6: 104 Chapter 36: 706–710

***Standards for Technological Literacy* Correlation Chart**	
The decision whether to develop a technology is influenced by societal opinions and demands, in addition to corporate cultures.	Chapter 2 Chapter 4: 61–63
A number of different factors, such as advertising, the strength of the economy, the goals of a company, and the latest fads, contribute to shaping the design of and demand for various technologies.	Chapter 4: 61–63, 75–76 Chapter 22: 429 Chapter 33: 664–665, 671 Chapter 35: 692, 700–702
Standard 7. Students will develop an understanding of the influence of technology on history.	Chapter 1: 17–20, 24–27 Chapter 2: 32–34 Chapter 4: 60–72, 75–76 Chapter 5: 97 Chapter 7: 127–142, 144–148 Chapter 8: 157–160, 165–168 Chapter 9: 187 Chapter 12: 224 Chapter 21: 413 Chapter 22: 374, 423–424 Chapter 23: 451 Chapter 25: 492 Chapter 27: 530 Chapter 28: 540–542 Chapter 29: 584 Chapter 33: 656–657 Chapter 35: 692–695 Chapter 36: 706–710
Benchmark Topics:	
Most technological development has been evolutionary, the result of a series of refinements to a basic invention.	Chapter 1: 17–18, 24–27 Chapter 4: 60 Chapter 8: 157–160 Chapter 33: 656–657
The evolution of civilization has been directly affected by, and has in turn affected, the development and use of tools and materials.	Chapter 1: 19, 24–27 Chapter 2: 32–34 Chapter 4: 63–72 Chapter 7: 127–139 Chapter 8: 165–168
Throughout history, technology has been a powerful force in reshaping the social, cultural, political, and economic landscape.	Chapter 1: 17–19, 24–27 Chapter 4: 60–63, 75–76 Chapter 35: 692–695
Early in the history of technology, the development of many tools and machines was based not on scientific knowledge but on technological know-how.	Chapter 1: 24 Chapter 22: 374

Standards for Technological Literacy Correlation Chart	
The Iron Age was defined by the use of iron and steel as the primary materials for tools.	Chapter 1: 24–25 Chapter 28: 540–542
The Middle Ages saw the development of many technological devices that produced long-lasting effects on technology and society.	Chapter 1: 25 Chapter 7: 146 Chapter 28: 542
The Renaissance, a time of rebirth of the arts and humanities, was also an important development in the history of technology.	Chapter 1: 26
The Industrial Revolution saw the development of continuous manufacturing, sophisticated transportation and communication systems, advanced construction practices, and improved education and leisure time.	Chapter 1: 26–27 Chapter 7: 134, 140–142, 144–148 Chapter 12: 224 Chapter 22: 423–424 Chapter 25: 492 Chapter 36: 708–709
The Information Age places emphasis on the processing and exchange of information.	Chapter 1: 27 Chapter 36: 709–710
Standard 8. Students will develop an understanding of the attributes of design.	Chapter 2: 36–37 Chapter 5: 82–84 Chapter 9: 179–187 Chapter 10: 192–202 Chapter 11 Chapter 12
Benchmark Topics:	
The design process includes defining a problem, brainstorming, researching and generating ideas, identifying criteria and specifying constraints, exploring possibilities, selecting an approach, developing a design proposal, making a model or prototype, testing and evaluating the design using specifications, refining the design, creating or making it, and communicating processes and results.	Chapter 2: 36–37 Chapter 5: 82–84 Chapter 9: 179–182 Chapter 10: 192–202 Chapter 11: 206–214 Chapter 12
Design problems are seldom presented in a clearly defined form.	Chapter 5: 82–83 Chapter 9: 183–187
The design needs to be continually checked and critiqued, and the ideas of the design must be redefined and improved.	Chapter 5: 83–84 Chapter 10: 195–196 Chapter 11: 214–218
Requirements of a design, such as criteria, constraints, and efficiency, sometimes compete with each other.	Chapter 5: 82–83 Chapter 10: 192

Standards for Technological Literacy Correlation Chart

Standard 9. Students will develop an understanding of engineering design.	Chapter 2: 36–37 Chapter 5: 82–84 Chapter 9 Chapter 10: 192 Chapter 11: 212–218 Chapter 12 Chapter 20
Benchmark Topics:	
Established design principles are used to evaluate existing designs, to collect data, and to guide the design process.	Chapter 12: 231 Chapter 20
Engineering design is influenced by personal characteristics, such as creativity, resourcefulness, and the ability to visualize and think abstractly.	Chapter 2: 36–37 Chapter 5: 82–84 Chapter 9 Chapter 12
A prototype is a working model used to test a design concept by making actual observations and necessary adjustments.	Chapter 5: 84 Chapter 11: 212–213
The process of engineering design takes into account a number of factors.	Chapter 11: 214–218
Standard 10. Students will develop an understanding of the role of troubleshooting, research and development, invention and innovation, and experimentation in problem solving.	Chapter 3: 46–48 Chapter 9 Chapter 33: 654–658
Benchmark Topics:	
Research and development is a specific problem-solving approach that is used intensively in business and industry to prepare devices and systems for the marketplace.	Chapter 3: 46–48 Chapter 33: 654–658
Technological problems must be researched before they can be solved.	Chapter 9: 179, 182, 185–187
Not all problems are technological, and not every problem can be solved using technology.	Chapter 9: 179
Many technological problems require a multidisciplinary approach.	Chapter 9

Standards for Technological Literacy Correlation Chart

Standard 11. Students will develop abilities to apply the design process.	Chapter 5: 82–84 Chapter 9: 182–187 Chapter 10 Chapter 12 Chapter 17: 242–243 Chapter 20: 396–399 Chapter 21: 416–417 Chapter 22: 438–440 Chapter 23: 460–461
Benchmark Topics:	
Identify the design problem to solve and decide whether or not to address it.	Chapter 5: 82–83 Chapter 9: 183–187 Chapter 10: 194
Identify criteria and constraints and determine how these will affect the design process.	Chapter 5: 82–83 Chapter 9: 182, 184–185
Refine a design by using prototypes and modeling to ensure quality, efficiency, and productivity of the final product.	Chapter 5: 83–84
Evaluate the design solution using conceptual, physical, and mathematical models at various intervals of the design process in order to check for proper design and to note areas where improvements are needed.	Chapter 5: 84
Develop and produce a product or system using a design process.	Chapter 10 Chapter 17: 242–243 Chapter 20: 396–399 Chapter 21: 416–417 Chapter 22: 438–440 Chapter 23: 460–461
Evaluate final solutions and communicate observation, processes, and results of the entire design process, using verbal, graphic, quantitative, virtual, and written means, in addition to three-dimensional models.	Chapter 5: 84 Chapter 12

Standards for Technological Literacy Correlation Chart

Standard 12. Students will develop the abilities to use and maintain technological products and systems.	Chapter 7 Chapter 11: 221–222 Chapter 12: 239–241 Chapter 13: 250–252 Chapter 18: 347–350 Chapter 20: 398–399 Chapter 23: 460–461 Chapter 26: 515–516 Chapter 31: 634–635 Chapter 34: 675–680
Benchmark Topics:	
Document processes and procedures and communicate them to different audiences using appropriate oral and written techniques.	Chapter 12: 239
Diagnose a system that is malfunctioning and use tools, materials, machines, and knowledge to repair it.	Chapter 13: 250–252 Chapter 18: 347–350 Chapter 34: 677–679
Troubleshoot, analyze, and maintain systems to ensure safe and proper function and precision.	Chapter 7 Chapter 18: 347–350 Chapter 26: 515–516
Operate systems so that they function in the way they were designed.	Chapter 7: 151–153, 155
Use computers and calculators to access, retrieve, organize, process, maintain, interpret, and evaluate data and information in order to communicate.	Chapter 11: 221–222 Chapter 12: 240–241 Chapter 20: 398–399 Chapter 23: 460–461 Chapter 31: 634–635
Standard 13. Students will develop the abilities to assess the impact of products and systems.	Chapter 18 Chapter 31: 634–635 Chapter 33: 680–682 Chapter 34 Chapter 35: 692–695
Benchmark Topics:	
Collect information and evaluate its quality.	Chapter 31: 634–635
Synthesize data, analyze trends, and draw conclusions regarding the effect of technology on the individual, society, and the environment.	Chapter 31: 634–635

Standards for Technological Literacy Correlation Chart

Use assessment techniques, such as trend analysis and experimentation to make decisions about the future development of technology.	Chapter 34 Chapter 35: 692–695
Design forecasting techniques to evaluate the results of altering natural systems.	Chapter 18 Chapter 34
Standard 14. Students will develop an understanding of and be able to select and use medical technologies.	Section 1: 15 Chapter 3: 52 Chapter 5: 95 Chapter 31
Benchmark Topics:	
Medical technologies include prevention and rehabilitation, vaccines and pharmaceuticals, medical and surgical procedures, genetic engineering, and the systems within which health is protected and maintained.	Chapter 5: 95 Chapter 31
Telemedicine reflects the convergence of technological advances in a number of fields, including medicine, telecommunications, virtual presence, computer engineering, informatics, artificial intelligence, robotics, materials science, and perceptual psychology.	Chapter 31
The sciences of biochemistry and molecular biology have made it possible to manipulate the genetic information found in living creatures.	Section 1: 15 Chapter 3: 52
Standard 15. Students will develop an understanding of and be able to select and use agricultural and related biotechnologies.	Section 1: 15 Chapter 1 Chapter 3: 52 Chapter 5: 85–86 Section 9: 565 Chapter 29 Chapter 30 Chapter 31: 620–631 Section 11: 689 Chapter 35
Benchmark Topics:	
Agriculture includes a combination of businesses that use a wide array of products and systems to produce, process, and distribute food, fiber, fuel, chemical, and other useful products.	Chapter 29: 567–586

Standards for Technological Literacy Correlation Chart	
Biotechnology has applications in such areas as agriculture, pharmaceuticals, food and beverages, medicine, energy, the environment, and genetic engineering.	Section 1: 15 Chapter 3: 52 Section 9: 565 Chapter 29: 586–587 Chapter 30 Chapter 31: 620–631
Conservation is the process of controlling soil erosion, reducing sediment in waterways, conserving water, and improving water quality.	Chapter 1 Chapter 35
The engineering design and management of agricultural systems require knowledge of artificial ecosystems and the effects of technological development on flora and fauna.	Section 11: 689
Standard 16. Students will develop an understanding of and be able to select and use energy and power technologies.	Chapter 4: 70, 74–75 Chapter 5: 91–92 Chapter 7: 139–145 Chapter 15: 285 Chapter 27 Chapter 28
Benchmark Topics:	
Energy cannot be created nor destroyed; however, it can be converted from one form to another.	Chapter 5: 92 Chapter 7: 139–145 Chapter 27: 531 Chapter 28: 539–555
Energy can be grouped into major forms: thermal, radiant, electrical, mechanical, chemical, nuclear, and others.	Chapter 4: 74–75 Chapter 5: 91–92 Chapter 15: 285 Chapter 27: 529–531
It is impossible to build an engine to perform work that does not exhaust thermal energy to the surroundings.	Chapter 27 Chapter 28
Energy resources can be renewable or nonrenewable.	Chapter 4: 70 Chapter 27: 532–534 Chapter 28
Power systems must have a source of energy, a process, and loads.	Chapter 5: 91–92 Chapter 28: 556–559

Standards for Technological Literacy Correlation Chart

Standard 17. Students will develop an understanding of and be able to select and use information and communication technologies.	Chapter 5: 86–88 Chapter 7: 145–151 Section 6: 361 Chapter 19 Chapter 20 Chapter 21 Chapter 22 Chapter 23
Benchmark Topics:	
Information and communication technologies include the inputs, processes, and outputs associated with sending and receiving information.	Chapter 5: 86–88 Chapter 7: 145–151
Information and communication systems allow information to be transferred from human to human, human to machine, machine to human, and machine to machine.	Chapter 19: 365–366
Information and communication systems can be used to inform, persuade, entertain, control, manage, and educate.	Chapter 19: 364 Chapter 22
Communication systems are made up of source, encoder, transmitter, receiver, decoder, storage, retrieval, and destination.	Chapter 5: 86–88 Chapter 19: 364–365 Chapter 22
There are many ways to communicate information, such as graphic and electronic means.	Chapter 5: 86–88 Section 6: 361 Chapter 19: 366–370 Chapter 20 Chapter 21: 403–413 Chapter 22: 420–434
Technological knowledge and processes are communicated using symbols, measurement, conventions, icons, graphic images, and languages that incorporate a variety of visual, auditory, and tactile stimuli.	Chapter 23
Standard 18. Students will develop an understanding of and be able to select and use transportation technologies.	Chapter 5: 95–98 Section 7: 465 Chapter 24 Chapter 25 Chapter 26
Benchmark Topics:	
Transportation plays a vital role in the operation of other technologies, such as manufacturing, construction, communication, health and safety, and agriculture.	Chapter 24: 467–468 Chapter 26

Standards for Technological Literacy Correlation Chart

Intermodalism is the use of different modes of transportation, such as highways, railways, and waterways as part of an interconnected system that can move people and goods easily from one mode to another.	Chapter 24
Transportation services and methods have led to a population that is regularly on the move.	Section 7: 465 Chapter 24 Chapter 26: 512–514
The design of intelligent and nonintelligent transportation systems depends on many processes and innovative techniques.	Chapter 24: 468–473 Chapter 25: 476–479 Chapter 26: 508–512, 517
Standard 19. Students will develop an understanding of and be able to select and use manufacturing technologies.	Chapter 1 Chapter 4 Chapter 5: 93–95 Chapter 7: 134 Chapter 12: 224 Chapter 13 Chapter 14 Chapter 15 Chapter 16 Chapter 18: 344–350 Chapter 33: 658–665 Chapter 36: 708–709
Benchmark Topics:	
Servicing keeps products in good operating condition.	Chapter 13: 250–252 Chapter 18: 347–350
Materials have different qualities and may be classified as natural, synthetic, or mixed.	Chapter 4 Chapter 14
Durable goods are designed to operate for a long period of time, while non-durable goods are designed to operate for a short period of time.	Chapter 13
Manufacturing systems may be classified into types, such as customized production, batch production, and continuous production.	Chapter 7: 134 Chapter 33: 658–663
The interchangeability of parts increases the effectiveness of manufacturing processes.	Chapter 1 Chapter 12: 224 Chapter 36: 708–709
Chemical technologies provide a means for humans to alter or modify materials and to produce chemical products.	Chapter 15

Standards for Technological Literacy Correlation Chart	
Marketing involves establishing a product's identity, conducting research on its potential, advertising it, distributing it, and selling it.	Chapter 33: 664–665
Standard 20. Students will develop an understanding of and be able to select and use construction technologies.	Chapter 5: 88–91 Chapter 13 Chapter 17 Chapter 18: 347–351 Chapter 34: 678
Benchmark Topics:	
Infrastructure is the underlying base or basic framework of a system.	Chapter 17
Structures are constructed using a variety of processes and procedures.	Chapter 5: 88–91 Chapter 34: 678
The design of structures includes a number of requirements.	Chapter 17
Structures require maintenance, alteration, or renovation periodically to improve them or to alter their intended use.	Chapter 5: 91 Chapter 13 Chapter 18: 347–351
Structures can include prefabricated materials.	Chapter 17

Basic Skills Chart

The Basic Skills Chart has been designed to identify those activities in *Technology* and the *Technology Student Activity Manual* that specifically encourage the development of basic skills.

Academic areas addressed in the chart include reading, writing, verbal (other than reading and writing), math, science, and analytical.

- **Reading** activities include assignments designed to improve comprehension of information presented in the chapter. Some are designed to improve understanding of vocabulary terms.
- **Writing** activities allow students to practice composition skills, such as letter writing, informative writing, and creative writing.
- **Verbal** activities encourage students to organize ideas, develop interpersonal and group speaking skills, and respond appropriately to verbal messages. Activities include oral reports and interviews.
- **Math** activities require students to use basic principles of math, as well as computation skills, to solve typical problems.
- **Science** activities call for students to use fundamental principles of science to solve typical problems.
- **Analytical** activities involve the higher-order skills needed for thinking creatively, making decisions, solving problems, visualizing information, reasoning, and knowing how to learn.

Activities are broken down by chapter, and a page number is given to locate the activity.

Basic Skills Chart	
Section 1	
Reading	**Text:** 1: Technology Defined (17); Technology As a Dynamic Process (19); The Presidential Election of 1960 (20); Positive and Negative Aspects of Technology (20); Technology and Types of Knowledge (21); Smart Houses (23); The Evolution of Technology (24) 2: System Components (31); Solar Collectors (39) 3: Level of Development (44); Economic Structure (46); Number of People Involved (48); Type of Technology Developed and Used (49)
Writing	**Text:** 2: Activity 2 (42)
Verbal	**Text:** 3: Activity 3 (55)
Math	
Science	**Text:** 1: Activity 2 (29); Activity 3 (29) 2: Newton's Third Law of Motion (34) 3: Genetic Engineering (52)
Analytical	**Text:** 1: Activity 1 (29); Activity 2 (29); Activity 3 (29) 2: Activity 1 (42) 3: Activity 1 (55); Activity 2 (55); Activity 3 (55) Activity 1A: Design Problem (56) Activity 1B: Fabrication Problem (56) **Student Activity Manual:** Activity 1A: Design Problem (7) Activity 1B: Fabrication Problem (9)

Basic Skills Chart	
Section 2	
Reading	**Text:** 4: People (61); Tools and Machines (63); Materials (69); Information (72); Energy (74); Finances (75); Time (76) 5: Problem-Solving or Design Processes (82); Production Processes (85); Hybrid Vehicles (90); The Tennessee Valley Authority (97); Management Processes (99) 6: Outputs: Desirable and Undesirable (105); Outputs: Intended and Unintended (107); Outputs: Immediate and Delayed (107); Feedback and Control (109)
Writing	
Verbal	
Math	**Text:** 4: The Law of Equilibrium (73)
Science	**Text:** 6: Integrated Circuits (108); Chlorofluorocarbons (117)
Analytical	**Text:** 4: Activity 1 (78); Activity 2 (78); Activity 3 (78) 5: Activity 1 (102); Activity 2 (102); Activity 3 (102) 6: Activity 1 (121); Activity 2 (121); Activity 3 (121) Activity 2A: Design Problem (122) Activity 2B: Fabrication Problem (122) **Student Activity Manual:** Activity 2A: Design Problem (13) Activity 2B: Fabrication Problem (15)

Basic Skills Chart	
Section 3	
Reading	**Text:** 7: Material-Processing Tools and Machines (127); Types of Machine Tools (131); Flexible Manufacturing (134); Energy-Processing Converters (139); Information-Processing Machines (145); Computer Bugs (148); Using Technology Safely (151) 8: Measurement Systems: Past and Present (157); Qualities Measured (160); Measurement Tools (165); Measurement and Control (168)
Writing	**Text:** 7: Activity 2 (155); Activity 4 (155); Activity 5 (155) Activity 3A: Design Problem (174) **Student Activity Manual:** Activity 3A: Design Problem (19)
Verbal	
Math	**Text:** 8: Types of Measurement (164); Measuring Area (171); Activity 1 (173); Activity 2 (173)
Science	
Analytical	**Text:** 7: Activity 1 (155); Activity 3 (155); Activity 4 (155); Activity 6 (155) 8: Activity 3 (173) Activity 3A: Design Problem (174) Activity 3B: Fabrication Problem (174) **Student Activity Manual:** Activity 3A: Design Problem (19) Activity 3B: Fabrication Problem (23) Taking Your Technology Knowledge Home: Designing a Tool (27)

Basic Skills Chart	
Section 4	
Reading	**Text:** 9: The Problem-Solving/Design Process in General (179); The Problem-Solving/Design Process in Technology (179); Steps in Solving Technological Problems and Meeting Opportunities (182); Identifying a Technological Problem or Opportunity (183); The Origin of Radar (187) 10: System Design (190); Product Design (192); Developing Design Solutions (192) 11: Modeling Design Solutions (206); Radar (216); Redesigning Products and Structures (218) 12: Product Documents and Reports (224); Principles of Design (231); Computer-Aided Design (234); Approval Documents and Reports (237)
Writing	**Text:** 9: Activity 1 (189) 11: Activity 2 (220) **Student Activity Manual:** Activity 4A/4B: Design/Fabrication Problem (33)
Verbal	
Math	**Text:** 10: Solid Geometry (198)
Science	
Analytical	**Text:** 9: Activity 1 (189); Activity 2 (189) 10: Activity 1 (204); Activity 2 (204); Activity 3 (204); Activity 4 (204) 11: Analyzing the Design (214); Activity 1 (220); Activity 2 (220); Computer-Aided Design, Engineering Animation (221) 12: Activity 1 (239); Activity 2 (239); Computer-Aided Design, 2D Architectural (240) Activity 4A: Design Problem (242) Activity 4B: Fabrication Problem (242) **Student Activity Manual:** Activity 4A/4B: Design/Fabrication Problem (33) Taking Your Technology Knowledge Home: Designing a Code (39)

Basic Skills Chart	
Section 5	
Reading	**Text:** 13: Production Activities (246); Servicing and Repairing Products and Structures (250) 14: Types of Natural Material Resources (254); Locating and Obtaining Natural Resources (258); Activity 3 (269) 15: Mechanical Processes (271); Thermal Processes (275); Chemical and Electrochemical Processes (284) 16: Types of Manufacturing Processes (290); Robots (298) 17: Types of Structures (318); Word Origins (333) 18: Selecting Technological Products (344); Installing Technological Products (346); Maintaining Technological Products (347); Repairing Technological Products (348); Altering Technological Products (350); Disposing of Technological Products (352)
Writing	**Text:** 14: Activity 3 (269)
Verbal	
Math	**Text:** 15: Calculating Board Footage (277)
Science	**Text:** 13: The Principles of Expansion and Contraction (251) 14: Synthetic Fuels (265) 15: Nuclear Energy (285) 18: Materials Science (353)
Analytical	**Text:** 13: Activity 1 (253) 14: Activity 1 (269); Activity 2 (269) 15: Activity 1 (288); Activity 2 (288); Activity 3 (288) 16: Activity 1 (317); Activity 2 (317) 17: Activity 1 (341); Activity 2 (341); Structural Engineering (342) 18: Activity 1 (355); Activity 2 (355) Activity 5A: Design Problem—Manufacturing Technology (356) Activity 5B: Fabrication Problem—Manufacturing Technology (356) Activity 5C: Design Problem—Construction Problem (358) Activity 5D: Fabrication Problem—Construction Problem (358) **Student Activity Manual:** Activity 5A/5B: Manufacturing Design/Fabrication Problem (45) Activity 5C: Construction Design Problem (53) Activity 5D: Construction Design Problem (55) Activity 5E: Product Servicing Activity (57) Activity 5F: Papermaking/Paper Recycling Activity (61) Taking Your Technology Knowledge Home: Designing a Manufactured Product (63)

Basic Skills Chart	
Section 6	
Reading	**Text:** 19: Items That Are Communicated (363); Goals of Communication (364); The Communication Model (364); Types of Communication (365); Communication Systems (366); Compact Disc (CD) Systems (368) 20: Printing Methods (375); Steps in Producing Printed Graphic Messages (378); Fax Machines (384); Computer-Based Publishing (390) 21: Fundamentals of Photographic Communication (403); Digital Theaters (405); Digital Photography (412); Other Types of Photographic Communication (412); The Beginnings of Photojournalism (413) 22: Types of Telecommunication Systems (420); Communicating with Telecommunication Systems (424); Fiber Optics (426); Advertising (429); Other Communication Technologies (434) 23: Computer Systems: An Overview (443); Networks (444); The Internet (445); Internet Access (446); Internet Domains (447); The World Wide Web (448); The Internet (451); Electronic Mail (452); Virtual Reality (454); Newsgroups and Chat Rooms (455); Electronic Commerce (456)
Writing	**Text:** 20: Desktop Publishing (398) 22: Activity 2 (437); Film Technology (438) Activity 6B: Production Problem (462) **Student Activity Manual:** Activity 6C: Telecommunication Activity (77) Taking Your Technology Knowledge Home: Promoting a Neighborhood Event (79)
Verbal	**Text:** 19: The Power of Radio (370) 22: Activity 3 (437); Film Technology (438) Activity 6B: Production Problem (462) **Student Activity Manual:** Activity 6C: Telecommunication Activity (77) Taking Your Technology Knowledge Home: Promoting a Neighborhood Event (79)
Math	**Text:** 20: Measuring Type (389)
Science	**Text:** 21: Light and Photography (401) 22: The Physics of Telecommunication (418); Film Technology (438)

(Continued)

Basic Skills Chart	
Section 6 *(continued from the previous page)*	
Analytical	**Text:** 19: Activity 1 (373); Activity 2 (373) 20: Activity 1 (395); Activity 2 (395); Promotional Graphics (396); Desktop Publishing (398) 21: Activity 1 (415); Activity 2 (415); Imaging Technology (416) 22: Activity 1 (437); Activity 2 (437); Activity 3 (437); Film Technology (438) 23: Activity 1 (459); Activity 2 (459); Cyberspace Pursuit (460) Activity 6A: Design Problem (462) Activity 6B: Production Problem (462) **Student Activity Manual:** Activity 6A: Design Problem (69) Activity 6B: Production Problem (73) Activity 6C: Telecommunication Activity (77) Taking Your Technology Knowledge Home: Promoting a Neighborhood Event (79)

Basic Skills Chart	
Section 7	
Reading	**Text:** 24: Transportation: A Definition (466); The Importance of Transportation (467); Transportation As a System (468); Types of Transportation Systems (468); Transportation System Components (469) 25: Vehicular Systems: An Overview (476); Land Transportation Vehicles (479); Water Transportation Vehicles (483); Maglev Trains (488); Air Transportation Vehicles (491); Space Transportation Vehicles (498) 26: Types of Transportation (506); Components of a Transportation System (508); Transporting People and Cargo (512); Maintaining Transportation Systems (515); Regulating Transportation Systems (517)
Writing	**Text:** 26: System Control Technology (520) Activity 7A: Design Problem (522) **Student Activity Manual:** Taking Your Technology Knowledge Home: Planning a Trip (89)
Verbal	
Math	**Text:** 25: Calculating Buoyant Force (501) 26: Relating Speed, Time, and Distance (515) **Student Activity Manual:** Taking Your Technology Knowledge Home: Planning a Trip (89)
Science	**Text:** 24: Newton's First Law of Motion (471)
Analytical	**Text:** 24: Activity 1 (475); Activity 2 (475) 25: Activity 1 (505); Activity 2 (505); Activity 3 (505) 26: Activity 1 (519); Activity 2 (519); Activity 3 (519); Activity 4 (519); System Control Technology (520) Activity 7A: Design Problem (522) Activity 7B: Fabrication Problem (522) **Student Activity Manual:** Activity 7A: Design Problem (83) Activity 7B: Fabrication Problem (85) Taking Your Technology Knowledge Home: Planning a Trip (89)

Basic Skills Chart	
Section 8	
Reading	**Text:** 27: Types of Energy (526); Energy, Work, and Power (527); Forms of Energy (529); Connections to Technology: History (530); Energy Is Interrelated (531); Sources of Energy (532); Energy Technology: Help or Harm (534) 28: Inexhaustible Energy Converters (539); Renewable Energy Converters (548); Thermal Energy Converters (549); Electrical Energy Converters (553); Applying Energy to Do Work (556)
Writing	**Text:** 27: Activity 2 (537) 28: Activity 1 (561); Activity 2 (561)
Verbal	
Math	**Text:** 27: Energy, Work, and Power (527); Origin of Horsepower (530) **Student Activity Manual:** Taking Your Technology Knowledge Home: Investigating Lighting Costs and Savings (101)
Science	**Text:** 27: Types of Energy (526); Energy Is Interrelated (531); Sources of Energy (532); Activity 1 (537) 28: Inexhaustible Energy Converters (539); Renewable Energy Converters (548); Thermal Energy Converters (549); Electrical Energy Converters (553); Laws of Gases (558); Activity 1 (561) Activity 8A: Design Problem (562) Activity 8B: Fabrication Problem (562)
Analytical	**Text:** 27: Activity 2 (537) 28: Activity 2 (561) **Student Activity Manual:** Activity 8A: Design Problem (95) Activity 8B: Fabrication Problem (99) Taking Your Technology Knowledge Home: Investigating Lighting Costs and Savings (101)

Basic Skills Chart	
Section 9	
Reading	**Text:** 29: Types of Agriculture (567); Crop Production (568); Technology in Crop Production (569); A Special Type of Crop Production (580); Raising Livestock (580); The Homestead Act and the Morrill Act (584); A Special Type of Animal Raising (585); Agriculture and Biotechnology (586) 30: Primary Food Processing (594); Secondary Food Processing (606) 31: Technology and Wellness (615); Technology and Illness (620); Dialysis Machines (624); Activity 1 (633)
Writing	**Text:** 29: Activity 1 (589); Activity 2 (589) 30: Activity 1 (613); Activity 2 (613) 31: Activity 1 (633); Activity 2 (633); Medical Technology (634) **Student Activity Manual:** Activity 9A: Design Problem (107)
Verbal	**Text:** 31: Medical Technology (634) **Student Activity Manual:** Activity 9A: Design Problem (107) Activity 9B: Fabrication Problem (111) Taking Your Technology Knowledge Home: Practicing Primary and Secondary Food Processing (113)
Math	
Science	**Text:** 29: Agriculture and Biotechnology (586) 30: Irradiation (607); Activity 1 (613); Activity 2 (613) 31: Aerodynamics (625) Activity 9B: Fabrication Problem (636) **Student Activity Manual:** Taking Your Technology Knowledge Home: Practicing Primary and Secondary Food Processing (113)
Analytical	**Text:** 29: Activity 2 (589) 31: Activity 1 (633) Activity 9A: Design Problem (636) **Student Activity Manual:** Activity 9A: Design Problem (107)

Basic Skills Chart	
Section 10	
Reading	**Text:** 32: Technology and the Entrepreneur (640); Technology and Management (641); Risks and Rewards (644); Forming a Company (645); Financing the Company (648) 33: Societal Institutions (652); Economic Enterprises (653); Industry (654); Areas of Industrial Activity (654); Research and Development (655); Production (658); Marketing (664); Industrial Relations (665); Financial Affairs (669); Industry-Consumer Product Cycle (671) 34: Using Technology (675); Assessing Technology (680)
Writing	**Text:** 32: Activity 2 (651) Activity 10A: Forming the Company (684) Activity 10B: Operating the Company (685) **Student Activity Manual:** Taking Your Technology Knowledge Home: Developing an Organization and Management System (133)
Verbal	**Text:** 32: Activity 2 (651) **Student Activity Manual:** Taking Your Technology Knowledge Home: Developing an Organization and Management System (133)
Math	**Text:** 33: Calculating Bids (668) Activity 10B: Operating the Company (685) **Student Activity Manual:** Activity 10B: Operating the Company (121)
Science	**Text:** 33: Wind Tunnels (660) 34: Earth-Sheltered Buildings (678)
Analytical	**Text:** 32: Activity 1 (651) 33: Activity 1 (673); Activity 2 (673) 34: Activity 1 (683); Activity 2 (683); Activity 3 (683) Activity 10A: Forming the Company (684) Activity 10B: Operating the Company (685) **Student Activity Manual:** Activity 10A: Forming the Company (119) Activity 10B: Operating the Company (121) Activity 10C: Using and Assessing Activity (131) Taking Your Technology Knowledge Home: Developing an Organization and Management System (133)

Basic Skills Chart	
Section 11	
Reading	**Text:** 35: Technology Controls and Harnesses Natural Forces (691); Technology Has a Global Impact (692); Technology and the Future (692); Mechanization (694); Technology: Challenges and Promises (695) 36: Technology and Lifestyle (706); Technology and Employment (711); Technology and Individual Control (715); Technology and Major Concerns (716); Technology and New Horizons (718); Activity 2 (721)
Writing	**Text:** 35: Activity 1 (704) 36: Activity 1 (721); Activity 2 (721) Activity 11A: Design Problem (722) **Student Activity Manual:** Activity 11B: Production Problem (139)
Verbal	**Text:** Activity 11A: Design Problem (722)
Math	
Science	35: Technology Controls and Harnesses Natural Forces (691)
Analytical	**Text:** 35: Activity 1 (704); Activity 2 (704) **Student Activity Manual:** Activity 11A: Design Problem (137) Activity 11B: Production Problem (139)

Scope and Sequence Chart

In planning your program, you may want to use this Scope and Sequence Chart to identify the major concepts presented in each chapter of *Technology*. Refer to the chart to select topics that meet your curriculum needs. The chart is divided into eleven sections. Within these sections, bold numbers indicate the chapters in which the concepts are found. Topics and their corresponding page numbers follow the chapter numbers for easy reference to the text.

Scope and Sequence Chart

Section 1

Core Concepts of Technology

1: Technology Defined (17); Technology As a Dynamic Process (19); Technology and Types of Knowledge (21)

Connections to Other Fields of Study

1: The Evolution of Technology (24); The Stone Age (24); The Bronze Age (24); The Iron Age (24); The Middle Ages (25); The Renaissance (26); The Industrial Revolution (26); The Information Age (27)

Technology and Its Impacts

1: Positive and Negative Aspects of Technology (20); The Evolution of Technology (24); The Stone Age (24); The Bronze Age (24); The Iron Age (24); The Middle Ages (25); The Renaissance (26); The Industrial Revolution (26); The Information Age (27)

Technological Systems and Processes

2: System Components (31); Goals (31); Inputs (32); Processes (36); Outputs (38); Feedback and Control (40)

3: Level of Development (44); Obsolete Technologies (45); Current Technologies (45); Emerging Technologies (45); Economic Structure (46); Profit-Centered Companies (46); Government (47); Number of People Involved (48); Type of Technology Developed and Used (49)

Section 2

Technology and Its Impacts

6: Outputs: Desirable and Undesirable (105); Outputs: Immediate and Delayed (107)

Technological Tools and Materials

4: Tools and Machines (63); Hand Tools (64); Machines (64); Materials (69); Types of Materials (69); Properties of Materials (72)

Scope and Sequence Chart

Technological Systems and Processes

4: People (61); Tools and Machines (63); Information (72); Energy (74); Types of Energy (74); Sources of Energy (74); Finances (75); Time (76)

5: Problem-Solving or Design Processes (82); Identifying the Problem or Opportunity (82); Developing Multiple Solutions (83); Isolating, Refining, and Detailing the Best Solution (83); Modeling and Evaluating the Selected Solution (84); Communicating the Final Solution (84); Production Processes (85); Agricultural and Related Biotechnical Processes (85); Communication and Information Processes (86); Construction Processes (88); Energy and Power Processes (91); Manufacturing Processes (93); Medical Processes (95); Transportation Processes (95); Management Processes (99)

6: Outputs: Desirable and Undesirable (105); Outputs: Immediate and Delayed (107); Feedback and Control (109); Internal Controls: Types of Control Systems (109); Internal Controls: Components of Control Systems (111); Internal Controls: The Operation (Manual versus Automatic) of Control Systems (114); External Controls (116)

Applications of Technology

5: Agricultural and Related Biotechnical Processes (85)

Technological Enterprises

4: Finances (75)

5: Management Processes (99)

Section 3

Design and Problem Solving in Technology

8: Measurement Systems: Past and Present (157); U.S. Customary System (157); SI Metric System (158); Qualities Measured (160); Size and Space (160); Mass (163); Temperature (163); Time (163); Types of Measurement (164); Standard Measurement (164); Precision Measurement (165); Measurement and Control (168); Measurement and Production Processes (168); Measurement and Quality Control (169)

Technological Tools and Materials

7: Material-Processing Tools and Machines (127); Characteristics of Machine Tools (127); The Cutting Tool (128); Motion (129); Support (130); Types of Machine Tools (131); Turning Machines (131); Sawing Machines (133); Drilling Machines (136); Shaping and Planing Machines (138); Grinding Machines (139); Energy-Processing Converters (139); The Steam Engine (140); The Internal Combustion Engine (142); The Electric Motor (144); Information-Processing Machines (145); Telecommunications: The Telephone (146); Telecommunications: The Radio (147); Computers (150); Using Technology Safely (151); Safety with People (151); Safety with Materials (152); Safety with Tools and Machines (152)

8: Measurement Tools (165); Direct-Reading Measurement Tools (165); Indirect-Reading Measurement Tools (168)

Technological Systems and Processes

Section 4

Design and Problem Solving in Technology

Section 5

Connections to Other Fields of Study

Technological Tools and Materials

Scope and Sequence Chart

Technological Systems and Processes

13: Production Activities (246); Resource Processing Systems (247); Production Manufacturing Systems (249); Structure Construction Systems (249)

15: Mechanical Processes (271); Producing Lumber (272); Thermal Processes (275); Producing Plywood (275); Producing Steel (277); Producing Glass (281); Refining Petroleum (282); Chemical and Electrochemical Processes (284)

16: Types of Manufacturing Processes (290); Casting and Molding Processes (291); Forming Processes (296); Separating Processes (301); Conditioning Processes (307); Assembling Processes (308); Finishing Processes (312)

Applications of Technology

13: Production Activities (246); Servicing and Repairing Products and Structures (250)

14: Locating and Obtaining Natural Resources (258)

15: Producing Lumber (272); Producing Plywood (275); Producing Steel (277); Producing Glass (281); Refining Petroleum (282)

17: Types of Structures (319); Buildings (319); Heavy Engineering Structures (335)

18: Selecting Technological Products (344); Installing Technological Products (346); Maintaining Technological Products (347); Repairing Technological Products (348); Altering Technological Products (350); Disposing of Technological Products (351)

Section 6

Connections to Other Fields of Study

21: Light and Photography (401)

22: The Physics of Telecommunication (418); Electrical Principles (419); Electromagnetic Waves (419)

Technology and Its Impacts

20: Designing the Message (378)

22: Designing the Message (424)

23: The World Wide Web (448)

Design and Problem Solving in Technology

21: Designing a Photographic Message (403)

22: Designing the Message (424)

Technological Systems and Processes

19: Items That Are Communicated (363); Goals of Communication (364); The Communication Model (364); Types of Communication (365); Communication Systems (366); Printed Graphic Communication (367); Photographic Communication (369); Telecommunication (369); Technical Graphic Communication (369); Computer and Internet Communication (370)

20: Computer-Based Publishing (390)

22: Types of Telecommunication Systems (420); Hard-Wired Systems (420); Broadcast Systems (422)

23: Computer Systems: An Overview (443)

Scope and Sequence Chart

Applications of Technology

20: Printing Methods (375); Relief Printing (375); Lithographic Printing (376); Gravure Printing (377); Screen Printing (377); Electrostatic Printing (378); Ink-Jet Printing (378); Steps in Producing Printed Graphic Messages (378); Designing the Message (378); Preparing to Produce the Message (381); Producing the Message (386)

21: Fundamentals of Photographic Communication; Designing a Photographic Message (403); Recording the Message (406); Reproducing the Message (410); Digital Photography (412); Other Types of Photographic Communication (412)

22: Communicating with Telecommunication Systems (424); Designing the Message (424); Producing Message (428); Delivering (Broadcasting) the Message (431); Other Communication Technologies (434); Mobile Communication Systems (434); Satellite Communication Systems (435); Sound Recording Systems (435)

23: Networks (444); Internet (445); Internet Access (446); Internet Domains (447); The World Wide Web (448); Web Browsers (449); Web Servers (449); Web Pages (449); Links (450); Search Engines (450); Electronic Mail (452); E-Mail System (452); E-Mail Messages (453); E-Mail Mailing Lists (455); Newsgroups and Chat Rooms (455)

Technological Enterprises

23: Electronic Commerce (456)

Section 7

Core Concepts of Technology

24: Transportation: A Definition (466); The Importance of Transportation (467)

Technological Systems and Processes

24: Transportation as a System (468); Transportation System Components (469); Pathways (470); Vehicles (472); Support Structures (472)

25: Vehicular Systems: An Overview (476); The Structural System (476); The Propulsion System (477); The Suspension System (477); The Guidance System (477); The Control System (478)

26: Components of a Transportation System (506); Transportation Routes (508); Transportation Schedules (511); Maintaining Transportation Systems (515); Regulating Transportation Systems (517)

Applications of Technology

24: Types of Transportation Systems (468); Land (468); Water (468); Air (468); Space (469)

25: Land Transportation Vehicles (479); Structure (479); Propulsion (480); Suspension (482); Guidance and Control (483); Water Transportation Vehicles (483); Structure (484); Propulsion (486); Suspension (487); Guidance and Control (490); Air Transportation Vehicles (491); Structure (492); Propulsion (493); Suspension (494); Guidance and Control (496); Space Transportation Vehicles (498); Types of Space Travel (499); Structure (500); Propulsion (502); Areas of Operation (502)

26: Types of Transportation (506); Components of a Transportation System (508); Vehicles (508); Transportation Terminals (511); Transporting People and Cargo (512); Loading (512); Moving (514); Unloading (514); Structure Repairs (515); Vehicle Repairs (516)

Section 8

Core Concepts of Technology

27: Types of Energy (526); Energy, Work, and Power (527); Forms of Energy (529)

Connections to Other Fields of Study

27: Renewable Energy Resources (532); Inexhaustible Energy Resources (533)

Technology and Its Impacts

27: Energy Technology: Help or Harm (534)

28: Renewable Energy Converters (548); Thermal Energy Converters (549); Electrical Energy Converters (553)

Technological Systems and Processes

28: Inexhaustible Energy Converters (539); Renewable Energy Converters (548); Mechanical Power Systems (556); Fluid Power Systems (558); Electrical Energy Converters (553)

Applications of Technology

27: Measuring Work (528); Measuring Power (528); Mechanical Energy (529); Radiant Energy (530); Chemical Energy (530); Thermal Energy (531); Electrical Energy (531); Nuclear Energy (531); Energy is Interrelated (531); Sources of Energy (532); Exhaustible Energy Resources (532); Renewable Energy Resources (532); Inexhaustible Energy Resources (533)

28: Inexhaustible Energy Converters (539); Wind Energy Conversion (540); Water Energy Conversion (542); Solar Energy Conversion (542); Geothermal Energy Conversion (546); Ocean Energy Conversion (547); Renewable Energy Converters (548); Thermochemical Conversion (548); Biochemical Conversion (548); Thermal Energy Converters (549); Heat Engines (549); Space Heating (550); Heat Production (551); Electrical Energy Converters (553); Generation (553); Distribution (555); Applying Energy to Do Work (556)

Scope and Sequence Chart

Section 9

Connections to Other Fields of Study

29: Agriculture and Biotechnology (586)

31: Goals of Medicine (621); Technology in Medicine (621)

Technology and Its Impacts

29: Agriculture and Biotechnology (586)

30: Food Preservation (601)

31: Technology and Wellness (615); Technology and Exercise (616); Technology and Sports (617); Technology and Illness (620); Goals of Medicine (621); Technology in Medicine (621)

Design and Problem Solving in Technology

29: Animal Waste Disposal Facilities (584)

Technological Tools and Materials

29: Power or Pulling Equipment (570); Planting Equipment (571); Pest Control Equipment (572); Irrigation Equipment (573); Harvesting Equipment (575); Transportation Equipment (578); Storage Equipment (579); Livestock Buildings (581); Fences and Fencing (582); Buildings and Machines for Feeding (583); Animal Waste Disposal Facilities (584)

31: Technology in Medicine (621)

Technological Systems and Processes

29: Types of Agriculture (567); Agriculture and Biotechnology (586)

30: Primary Food Processing (594); Secondary Food Processing (606)

Applications of Technology

29: Crop Production (568); Technology in Crop Production (570); A Special Type of Crop Production (580); Raising Livestock (580); Animal Waste Disposal Facilities (584); A Special Type of Animal Raising (585)

30: Material Conversion (594); Food Preservation (601); Developing Food Products (606); Manufacturing Food Products (608); An Example of Secondary Food Processing: Pasta Making (609)

31: Technology and Illness (620); Goals of Medicine (621); Technology in Medicine (621)

Section 10

Technology and Its Impacts

33: Research and Development (655)

34: Using Technology (675); Selecting Appropriate Devices and Systems (676); Selecting and Using Technological Services (680); Assessing Technology (680)

Scope and Sequence Chart

Applications of Technology

33: Production (658)

34: Selecting Appropriate Devices and Systems (676); Operating Products and Systems (676); Servicing and Repairing Devices and Systems (677); Disposing of Devices and Systems (679)

Technological Enterprises

32: Technology and the Entrepreneur (640); Technology and Management (641); Functions of Management (642); Authority and Responsibility (643); Risks and Rewards (644); Forming a Company (645); Selecting a Type of Ownership (645); Establishing the Enterprise (647); Electing a Board of Directors (648); Financing the Company (648); Equity Financing (648); Debt Financing (649)

33: Societal Institutions (652); Economic Enterprises (653); Industry (654); Areas of Industrial Activity (654); Research and Development (655); Research (655); Development (656); Engineering (657); Production (658); Planning (661); Producing (661); Maintaining Quality (661); Marketing (664); Industrial Relations (665); Employee Relations (666); Labor Relations (668); Public Relations (669); Financial Affairs (669); Industry-Consumer Product Cycle (671)

Section 11

Technology and Its Impacts

35: Technology Controls and Harnesses Natural Forces (691); Technology Has a Global Impact (692); Technology and the Future (692); Technology: Challenges and Promises (695); Energy Use (695); Environment Protection (696); Global Economic Competition (700)

36: Technology and Lifestyle (706); Colonial Life and Technology (706); The Industrial Revolution (708); The Information Age (709); Technology and Individual Control (715); Technology and Major Concerns (716); Technology and New Horizons (718)

Technological Enterprises

35: Global Economic Competition (700)

36: The Industrial Revolution (708); The Information Age (709); Technology and Employment (711); Types of Technical Jobs (711); Selecting a Job (712)

Section 1

Technology

Teaching Materials

Text

Student Activity Manual

Teacher's Resources

Transparencies (Binder/CD only)
1-1: Technology
1-2: Technology and Tools
2-1: System Components
3-1: Technology Is Constantly Changing

Organization of Section 1

In any educational program, students should understand the scope and sequence of the course they are taking. It is also a good practice to introduce the students to your operational procedures and practices early in the course. Therefore, Section 1 is designed to address three major topics:

- The structure (scope and sequence) of the course
- The hands-on nature of the study
- Technology as an intellectual discipline and as a mode of addressing human problems and needs

Specifically, this section allows students to develop an understanding of the role of technology in their lives and society. As shown in Figure 1-1, they will be introduced to the ideas that:

- Technology is human knowledge.
- Technology involves the use of tools, materials, and systems.

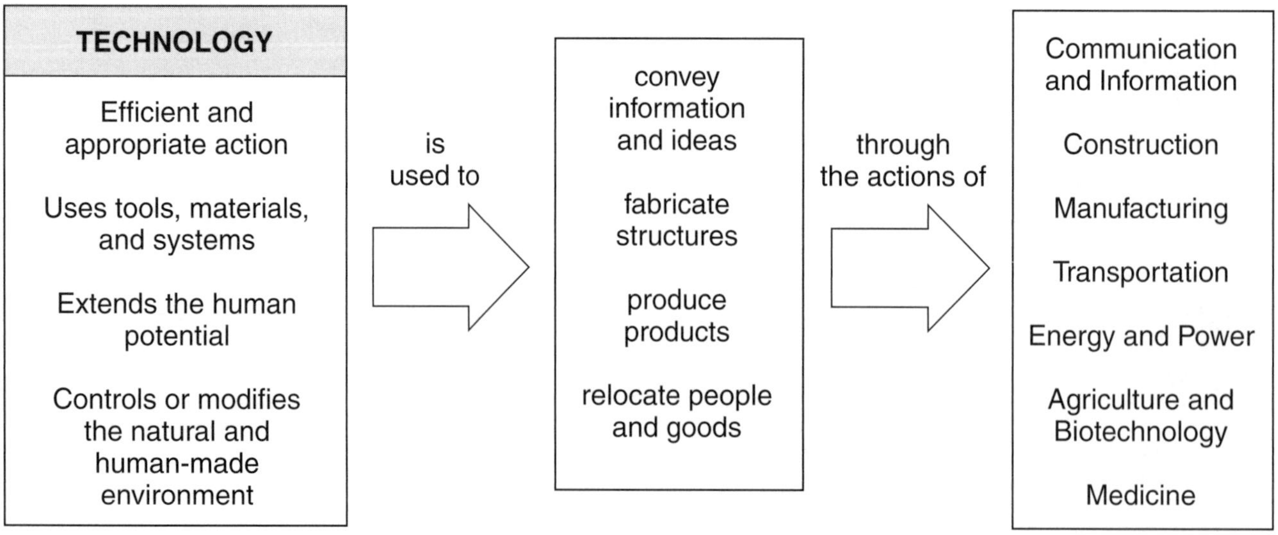

Figure 1-1

- Technology results in artifacts (physical objects) and other outputs (pollution and scrap, for example).
- Technology is developed by people to modify or control the environment.

Then, technology is contrasted with other types of human knowledge, Figure 1-2. Students will learn that people must use three types of knowledge to understand the world in which they live and to solve problems that arise. These types of knowledge are:

- Science: the knowledge of the natural world
- Technology: the knowledge of the human-made world
- Humanities: the knowledge of human values and interactions

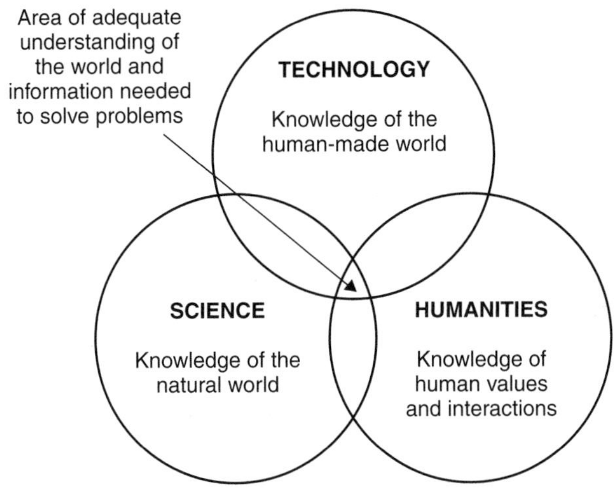

Figure 1-2

From there, the students explore technology as a human-designed and human-operated system. The students examine the major parts of all systems and see how they interrelate, Figure 1-3. This study includes:

Inputs—The elements that flow into the system and are consumed or processed by the system

- People
- Materials
- Tools and machines
- Energy
- Information

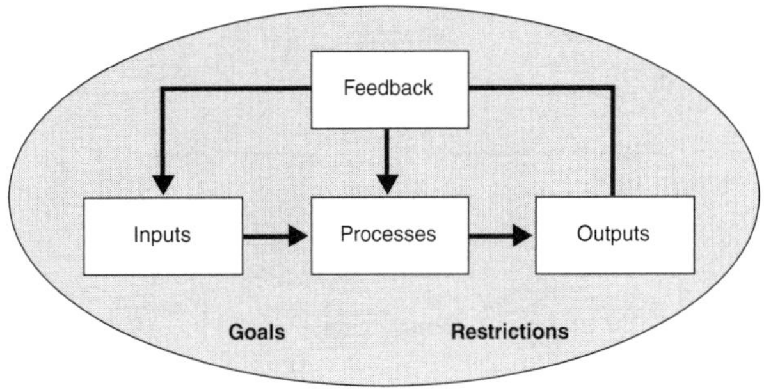

Figure 1-3

- Finances
- Time

Processes—Actions used to design, produce, and deliver goods and services

- Problem-solving/design processes
- Production processes
- Management processes

Outputs—The results of operating the system

- Intended and unintended
- Desirable and undesirable
- Immediate and delayed

Feedback and Control—Using information about the outputs of a process or system to regulate the system

Finally, the students see how technology is applied to economic activity. They are introduced to the major actions that change an idea into a product or service that is sold for a profit, Figure 1-4.

- Research and development: Designing, developing, and specifying the characteristics of a product, structure, or service
- Production: Developing and operating systems for producing the product, structure, or service
- Marketing: Promoting, selling, and delivering a product, structure, or service
- Industrial relations: Developing and administrating programs to ensure that the relations between the company and its workers and the public are positive
- Financial affairs: Obtaining the capital and physical resources and maintaining the financial records to manage the system properly

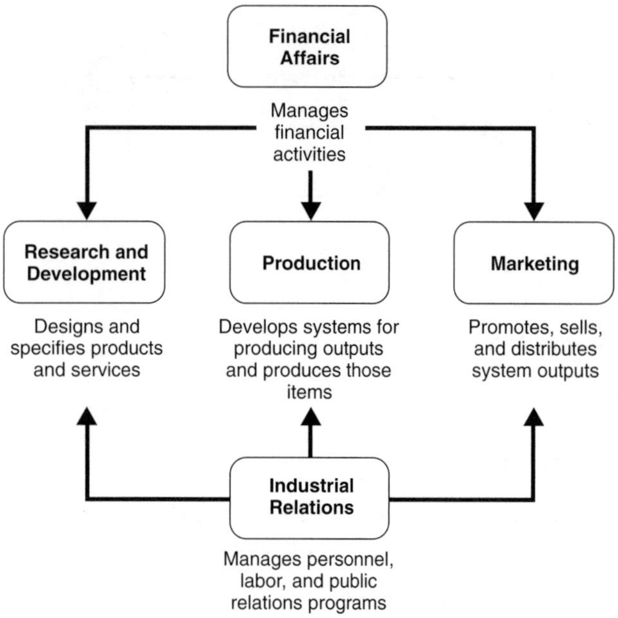

Figure 1-4

This section includes three chapters and one laboratory activity. The chapters are Chapter 1—Technology: A Dynamic, Human System, Chapter 2—Technology As a System, and Chapter 3—Types of Technological Systems.

As with each section, two different types of activities are provided. Those teachers wanting students to address the content from a problem-solving approach should select Activity 1A—Design Problem. A more structured laboratory activity is presented by Activity 1B—Fabrication Problem.

Presenting Section 1

Estimated length—1 week

Section 1 is designed to help students understand technology as:

- A body of knowledge of efficient and appropriate action
- A system with inputs, processes, outputs, feedback and control, and goals
- A group of practices that includes agricultural and related biotechnologies, communication and information technologies, construction technologies, energy and power technologies, manufacturing technologies, medical technologies, and transportation technologies.

The content of Section 1 lends itself to lecture/discussion strategies. It is more important, however, to use presentations that are reinforced with practical laboratory activities. This approach is appropriate for two reasons. First, students learn abstract concepts best when they are integrated with problem solving, critical thinking, and laboratory practice. Second, it is important to send the students a message that says, "This class is a fun, hands-on class where I can learn a great deal!"

Thus, it is important, starting with the first or second day, that all lectures/presentations be limited to only 10 to 15 minutes. All class periods need to have a large portion dedicated to laboratory activity. For this course, a laboratory activity is any experience in which students are actively involved in seeking solutions to a problem or any activity in which the students are engaged with tools, materials, or technological systems. This could involve designing products, brainstorming ideas, making sketches and drawings, completing activity sheets, building prototypes, or evaluating procedures.

A five-day schedule, shown in Figure 1-5, has been developed for this section. During this period of time, you will notice that three important topics are introduced through presentations and discussions. In addition, the students use simple tools to complete either a design or a fabrication problem. **Note:** It is important that the students use tools and machines SAFELY. Demonstrate appropriate safety procedures to your class prior to any laboratory activity.

To prepare to present this section, you should:

1. Read Chapters 1, 2, and 3 in *Technology*.
2. Prepare discussions on:
   a. What is technology? (Chapter 1)
   b. What are technological systems and their parts? (Chapter 2)
   c. What are the seven main types of technological systems? (Chapter 3)
3. Select a laboratory activity.
   a. Activity 1A—This is an open-ended problem-solving activity.
   b. Activity 1B—This is a structured activity in which groups of students use one of two techniques to produce

a simple product. They then compare the process without technology to the process that uses technology.

4. Practice the selected laboratory activity.
   a. Prepare teaching materials (transparencies or bulletin boards, for example) to support the content.
   b. Determine which study questions and test questions will be used.

Section 1: Technology	
Day	Activity
1	Complete "Start of the School Year" administrative details Discussion: The Technology Systems Course (Scope and sequence)
2	Discussion: "What Is Technology?" (Chapter 1) Introduce laboratory activity
3	Discussion: "What Is a Technological System?" (Chapter 2) Students work on laboratory activity
4	Students work on laboratory activity
5	Discussion: "What Are the Types of Technological Systems?" (Chapter 3) Students complete laboratory activity

Figure 1-5

Presenting the Activities

Consider the following recommendations in presenting the activity you select for this section.

Activity 1A—Design Problem

The students will be challenged to use technology (a technical device) to make a job easier and more efficient. This will reinforce the concepts of technology and technological systems, the focus of this section. To implement this activity:

1. Provide an introduction to the activity and discuss the background to Activity 1 presented in *Technology.*
2. Divide the class into groups of two to four students.
3. Present each group with the problem. Easy-Play Game Company has developed a new two-person board game that uses five red marbles and five green marbles as the game pieces. They find that counting the marbles individually is too costly.
4. Present each group with their challenge. Design a technological device (machine) that will automatically count five marbles at a time from a box containing a large number of marbles.
5. Have each member of each group:
   a. Work independently to develop at least two solutions to the problem.
   b. Present his/her solution to the group.
6. Have each group:
   a. Discuss each solution.
   b. Select a "best" solution.
   c. Construct a prototype (working model) of the solution.
   d. Test and evaluate the prototype.
   e. Develop a list of suggestions for improvement based on information gathered from the test.
   f. Present their solution to the class.

Activity 1 B—Fabrication Problem

1. Provide an introduction to the activity and discuss the background to Activity 1 presented in *Technology.*
2. Divide your class into two groups. The groups will use two different early clay-forming techniques to produce a clay pot and lid.
3. Demonstrate the production techniques. (See procedure on pages 56–57 in the textbook.)
4. Have group one form the two parts without the aid of any external devices. They will use the procedure in the Section Activity in the textbook. Group two will use a form (large frozen orange juice concentrate can) to aid in the product production process. They should

use the second procedure in the Section Activity in the textbook.

5. Meet as a class and analyze the two clay-forming techniques by answering the following questions:
   a. Do you think the coil-forming method is a common technique for producing modern ceramic products? Why or why not?
   b. Which method was easier to use?
   c. Which method produced the higher-quality products?

Reinforcing and Enriching

The following activities can be used to supplement and enrich the laboratory activities at the end of each textbook section and the Applying Your Knowledge activities included at the end of each textbook chapter.

1. Research early pottery making in Pueblo Indian cultures. Prepare either a report or a display to communicate the information gained from the research.
2. Select a simple device (tool or machine, for example). Make a bulletin board that shows the device as a technological system with inputs, processes, and outputs.

Adjustments for Exceptional Students

Both laboratory activities are designed to be done rather quickly. The level of sophistication can accommodate students of all ability levels. Gifted students should have more creative answers and may produce a more complex solution. Lower achievers can still solve the problem but with solutions within their ability level. In all cases, the amount of structure and teacher guidance should be adjusted to meet students' needs. Guard against providing too much structure, however, because this action often thwarts creativity and is less appealing to gifted students. In addition, careful grouping of students allows each student to bring his or her own special abilities to the group. Some students have more artistic abilities, others may have more dexterity, and still other students have more abstract reasoning ability. Throughout the course, group work allows students to develop cooperative work habits and to contribute their own unique abilities to the team.

Evaluation Procedures

The understanding that students have of technology and technological systems should be measured in several ways including:

- Observation and critical listening during class presentations and discussions.
- Worksheets and quizzes covering the cognitive material presented.
- Open-ended questions about the learned material and how technology is developed.
- Assessment of individual participation and contributions to group work.
- Assessment of the quality and creativeness of the solutions to the assigned design/fabrication problems.

Chapter 1
Technology: A Dynamic, Human System

Objectives

After studying Chapter 1, students will be able to:

- Define technology.
- List the basic features of technology.
- Describe technology as a dynamic process.
- List positive and negative features of technology.
- Explain the difference between science and technology.
- Name the major divisions in the evolution of technology.
- Cite technological developments in each period of technological history.
- Identify characteristics of the Information Age.

Answers to Test Your Knowledge *Questions, Text Pages 28–29*

1. Technology is humans using objects (tools, machines, systems, and materials) to change the natural and human-made (built) environment.
2. c. Technology is found in nature.
3. True
4. True
5. Evaluate individually.
6. Science
7. Science, technology, and the humanities
8. f. Industrial Revolution
9. d. Middle Ages
10. a. Stone Age
11. g. Information Age
12. b. Bronze Age
13. e. Renaissance
14. c. Iron Age
15. d. Middle Ages
16. a. Stone Age
17. e. Renaissance
18. f. Industrial Revolution
19. Steam engine
20. True

Answers to the Chapter Quiz, Teacher's Resource

1. False
2. True
3. True
4. True
5. False
6. False
7. True
8. True
9. False
10. c. Bronze Age
11. d. It can cause workers to lose their jobs.
12. a. Science is knowledge of the natural world, and technology is the use of tools to change natural and human-made environments.
13. Technology is humans using objects (tools, machines, systems, and materials) to change the natural and human-made (built) environment.
14. Any positive use of technology is a correct answer.
15. Any negative impact of technology, such as pollution or unemployment, is a correct answer.

Caption for Reproducible Master

Reproducible Master 1-1. Technology Can Help or Harm—Shown are one positive and one negative effect of technology.

Captions for Transparency Masters (Binder/CD Only)

Transparency Master 1-1. Technology—Technology is humans using objects to change the natural and human-made environment.

Transparency Master 1-2. Technology and Tools—Technology is any tool that extends the human potential to accomplish tasks.

Technology: A Dynamic, Human System

Name ______________________________ Period ______________

Date ______________________________ Score ______________

Chapter 1 Quiz

True/False: If a statement is true, write *True* in the blank. If a statement is false, write *False* in the blank.

_________ 1. Technology is knowledge of the natural world.

_________ 2. During the Industrial Revolution, large numbers of people left farms and migrated to growing towns and cities.

_________ 3. Technology is always changing and causing change.

_________ 4. Technology extends human abilities or potential to do work.

_________ 5. Only companies within the same country are in competition.

_________ 6. There was no technological advancement during the Dark Ages.

_________ 7. Nearly all technology is developed and produced by profit-centered businesses.

_________ 8. Humans have been distinguished from other species by their ability to make and use tools.

_________ 9. The Middle Ages immediately followed the Renaissance.

Multiple Choice: Choose the best answer and write the corresponding letter in the blank.

_____ 10. Large-scale irrigation systems were developed during which of the following time periods?
- a. Iron Age
- b. Stone Age
- c. Bronze Age
- d. Industrial Revolution

_____ 11. Which of the following is a downside of technology?
- a. It increases productivity.
- b. Technological advancement weakens the economy.
- c. It increases product costs.
- d. It can cause workers to lose their jobs.

_____ 12. Which of the following is an accurate description of the difference between science and technology?
- a. Science is knowledge of the natural world, and technology is the use of tools to change natural and human-made environments.
- b. Science is a comprehensive study of the universe, and technology is based on only the most recent scientific discoveries.
- c. Science is a form of knowledge, and technology is not.
- d. Science and technology are completely unrelated.

(Continued)

Short Answer: In the space provided, write a brief answer to each of the following questions.

13. Define *technology*.

14. Describe two ways technology has made your life better.

15. Describe two ways technology has had a negative effect on you or the world around you.

Technology can help or harm.

The difference is how people develop and use it.

Chapter 2

Technology As a System

Objectives

After studying Chapter 2, students will be able to:

- Explain how technology is a system.
- Identify the major components of a technological system.
- List the inputs to a technological system.
- List the steps in the problem-solving/design process.
- Cite the major activities involved in production processes.
- Cite the major activities involved in management processes.
- Name positive and negative technological outputs.
- Explain feedback and control.

Answers to Test Your Knowledge *Questions, Text Pages 41–42*

1. Each technology has parts, and each part has a relationship with all other parts and to the whole. The parts work together in a predictable way, and they are designed to achieve a goal.
2. Goals, inputs, processes, outputs, feedback and control
3. False
4. People, materials, tools and machines, energy, information, finances, and time
5. People
6. True
7. True
8. Information is data that has been organized and grouped according to its type.
9. False
10. Identifying the problem. Developing solutions to the problem. Isolating, refining, and detailing the best solution. Modeling and evaluating the solution. Communicating the final solution.
11. Planning, organizing, actuating, and controlling
12. Evaluate individually.
13. Evaluate individually.
14. Evaluate individually.
15. Feedback and control

Answers to the Chapter Quiz, Teacher's Resource

1. True
2. False
3. False
4. True
5. False
6. True
7. True
8. False
9. c. Inputs
10. a. Modeling and evaluating the solution
11. b. Production processes
12. c. Organizing
13. Goals, inputs, processes, outputs, and feedback and control
14. Data are facts and figures. Information is organized data. Knowledge is the use of information to understand, interpret, or describe a specific situation or series of events.
15. Feedback and control is the use of information about the outputs of a process or system to regulate the system.

Captions for Reproducible Masters

Reproducible Master 2-1. Technology As a System—Shown are some of the things that technology is and does.

Reproducible Master 2-2. Technological Processes—Shown are the seven inputs to technology systems.

Reproducible Master 2-3. Technological Processes—Shown is the problem-solving/design process, which is one of the three important processes that take place in technology systems.

Reproducible Master 2-4. Technological Processes—Shown are the seven types of production processes that take place in technology systems.

Reproducible Master 2-5. Technological Processes—Shown is the management process, which is one of the three of the important processes that take place in technology systems.

Reproducible Master 2-6. Technological Processes—Shown is a selection of outputs for a technological system.

Caption for Transparency Master (Binder/CD Only)

Transparency Master 2-1. System Components—Shown are the five important components of a system.

Technology As a System

Name ______________________________ Period ______________

Date ______________________________ Score ______________

Chapter 2 Quiz

True/False: If a statement is true, write *True* in the blank. If a statement is false, write *False* in the blank.

_________ 1. All technological developments have been responses to human needs.

_________ 2. Tools are the most important input to technological systems.

_________ 3. Information is the use of organized data to understand or interpret a specific situation.

_________ 4. Management processes direct the design, development, production, and marketing of a technological device, service, or system.

_________ 5. Outputs are used to regulate a system.

_________ 6. People make technological systems function and consume the products and services provided by the technological systems.

_________ 7. Synthetic materials are made by humans.

_________ 8. Machines can be defined as tools that contain more than five parts.

Multiple Choice: Choose the best answer and write the corresponding letter in the blank.

_____ 9. Which of the following is consumed or processed by a technological system?
a. Goals
b. Feedback
c. Inputs
d. Outputs

_____ 10. Which step of the problem-solving process involves producing physical, graphic, or mathematical models?
a. Modeling and evaluating the solution
b. Developing solutions to the problem
c. Isolating and detailing the best solution
d. Communicating the final solution

_____ 11. Which of the following are actions that are completed to perform the function of the technological system?
a. Problem-solving processes
b. Production processes
c. Management processes
d. Controlling

_____ 12. Which of the following involves dividing tasks into major segments and assigning resources to each task?
a. Actuating
b. Controlling
c. Organizing
d. Planning

(Continued)

Name ______________________________

Short Answer: In the space provided, write a brief answer to each of the following questions.

13. List the five major components present in all systems.

14. Describe the difference between data, information, and knowledge.

15. Explain feedback and control.

Technology As a System

Comes from a human need.

Is designed and developed by people.

Uses resources to produce outputs.

Is evaluated by people.

Over time, is modified or abandoned.

Has impacts on people, society, and the environment.

Technological Processes

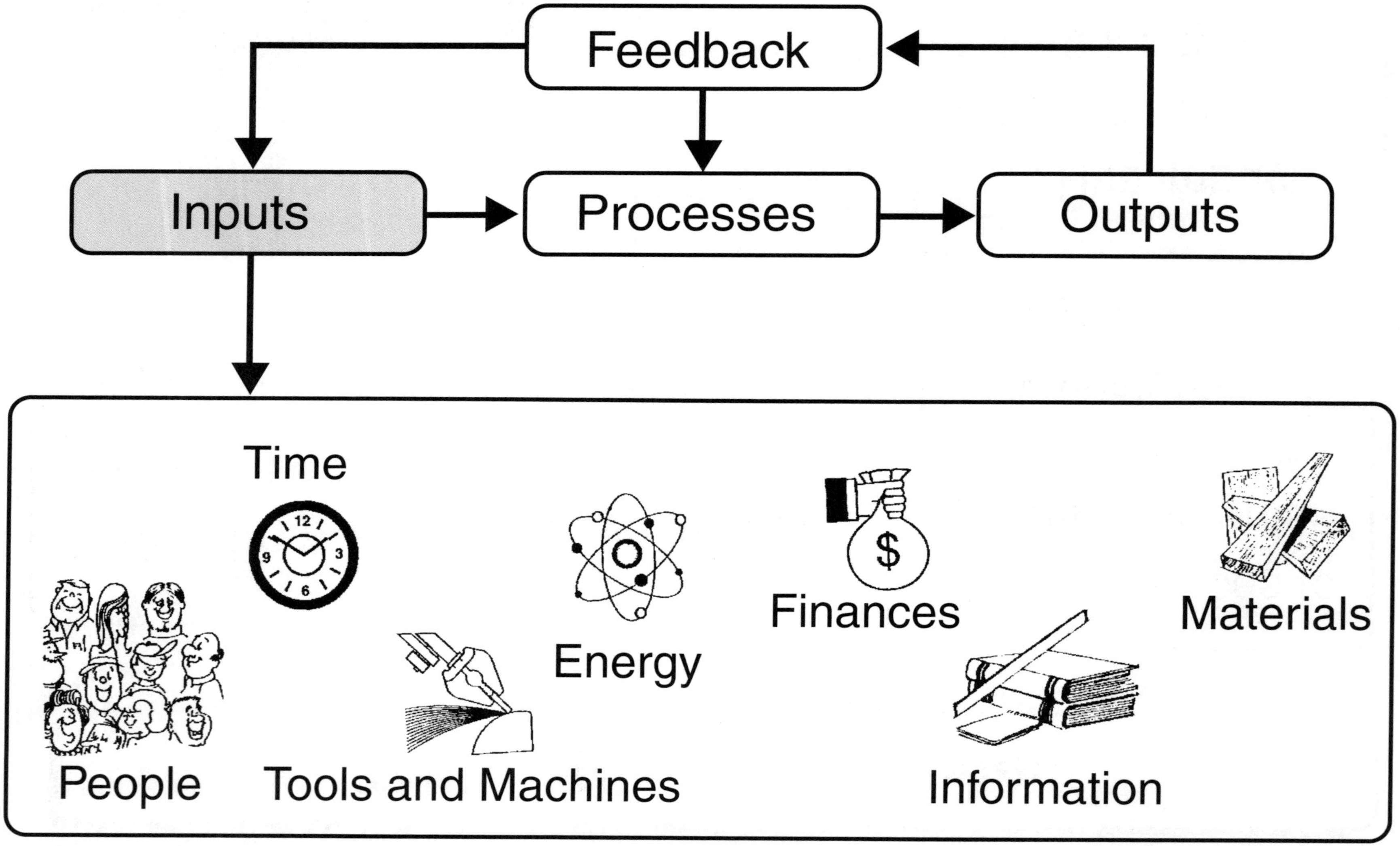

Technological Processes

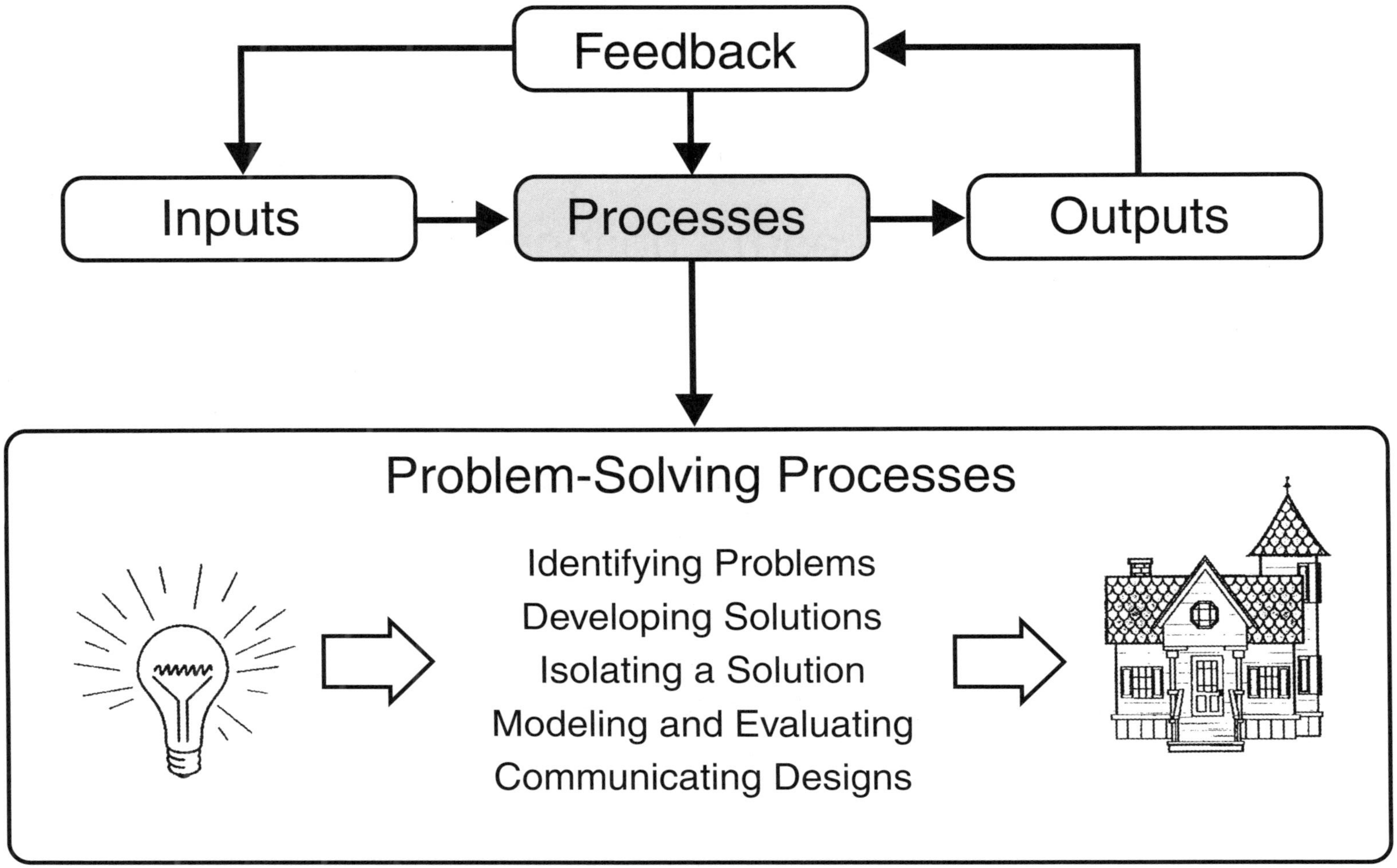

Technological Processes

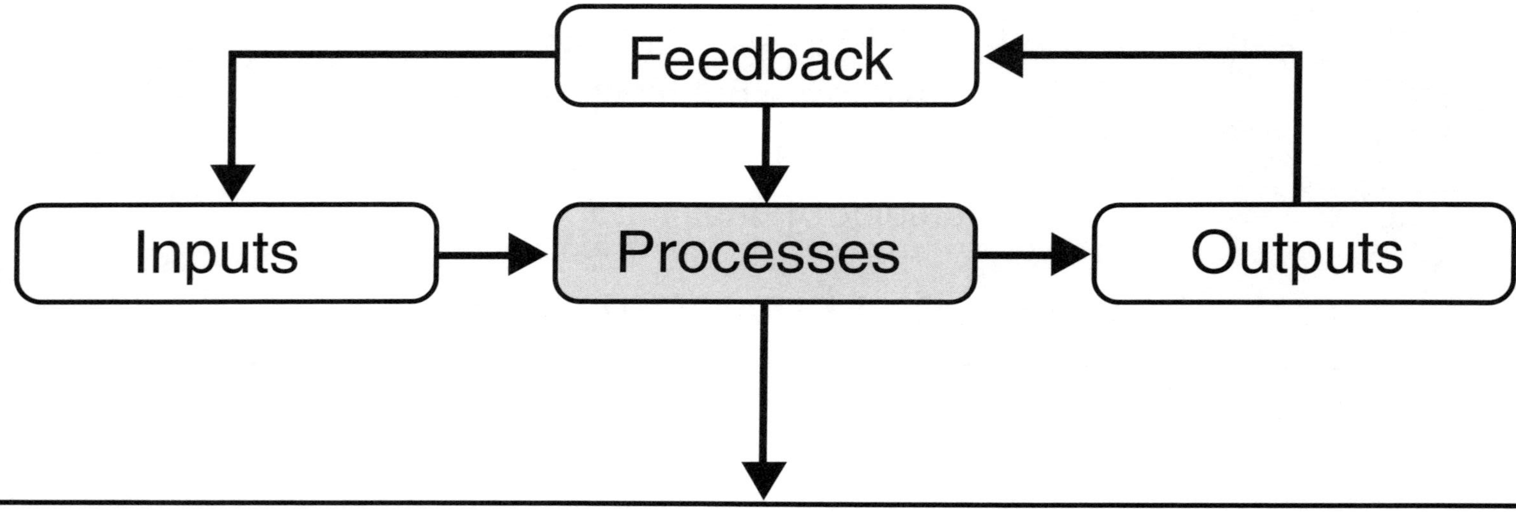

Production Processes

Agriculture Processes

Construction Processes

Information and Communication Processes

Energy Conversion Processes

Manufacturing Processes

Medical Processes

Transportation Processes

Technological Processes

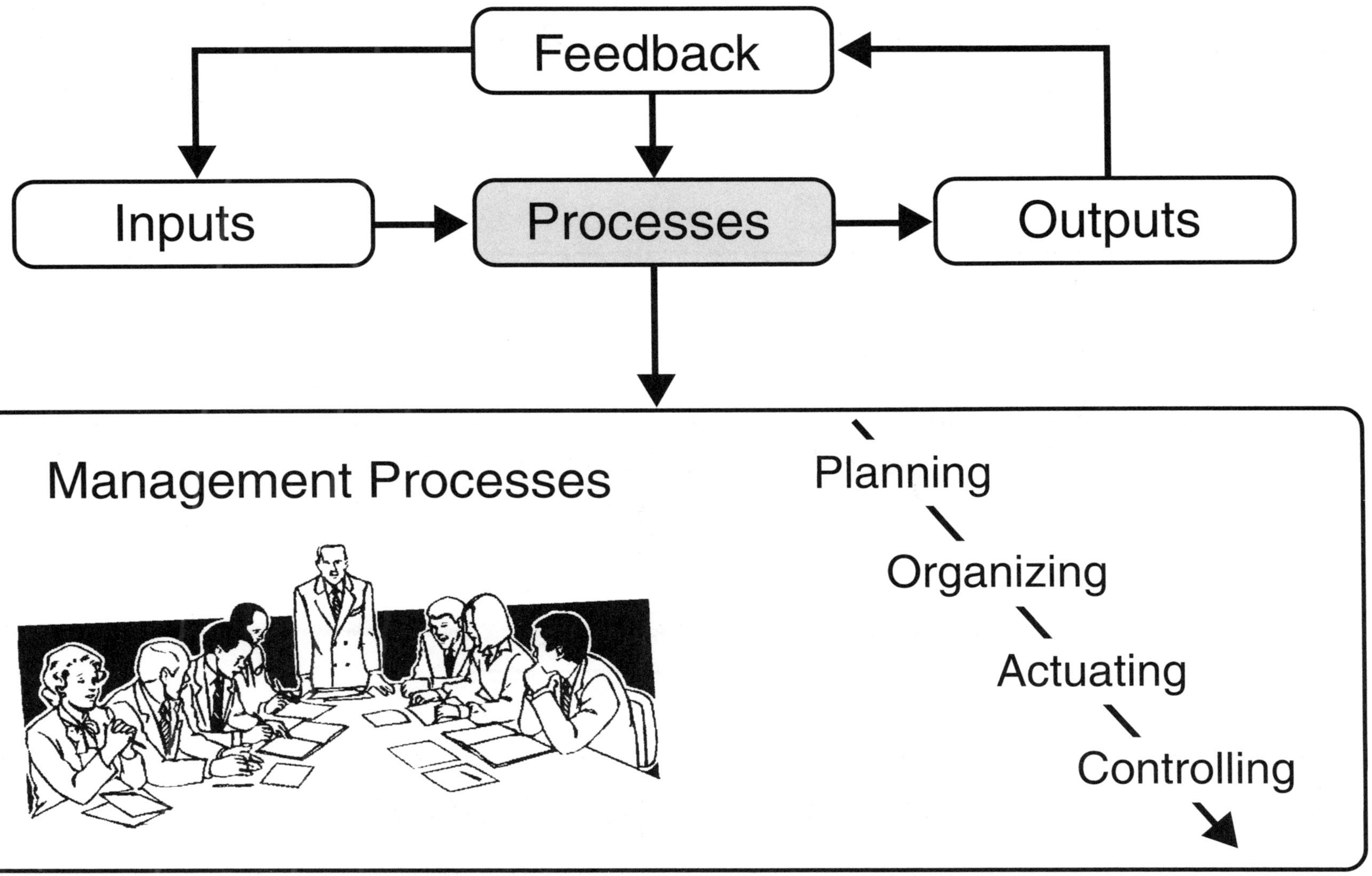

Technological Processes

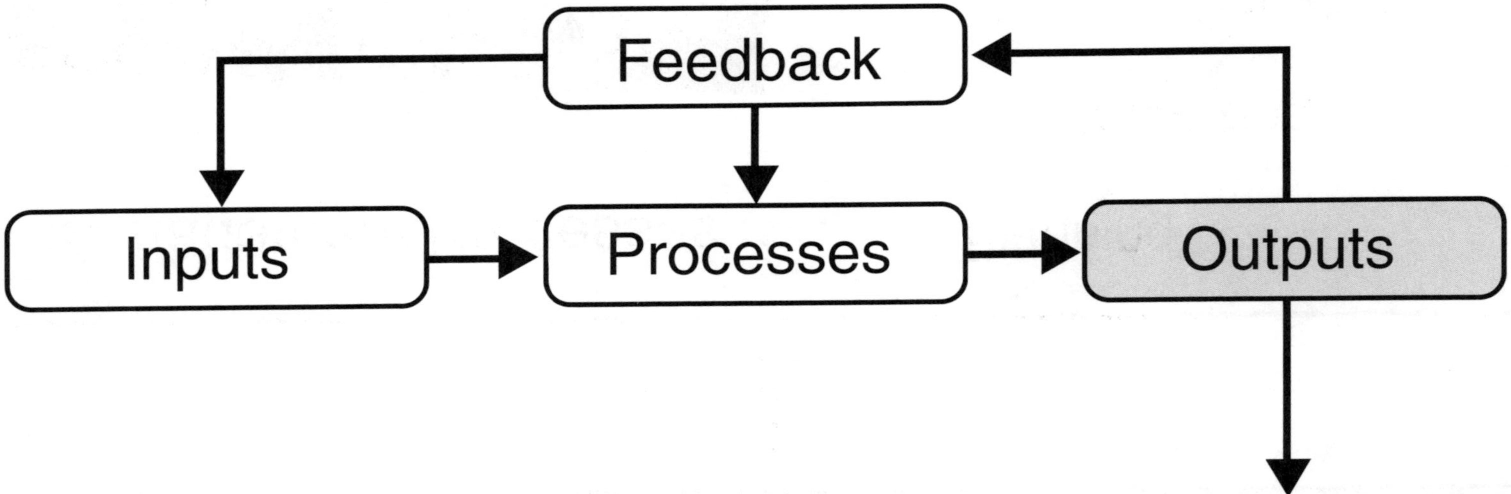

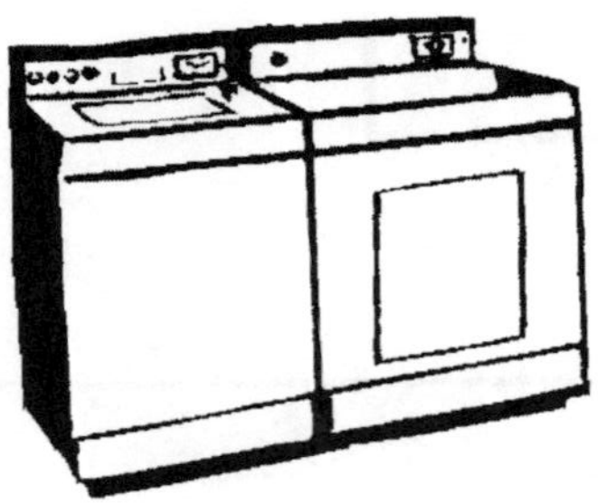

Desired or Undesired

Intended or Unintended

Immediate or Delayed

Chapter 3

Types of Technological Systems

Objectives

After studying Chapter 3, students will be able to:

- List the four general ways technology systems can be categorized.
- Identify obsolete, current, and emerging technologies.
- Describe how technology evolves in a private enterprise system.
- Cite examples of government-sponsored technology.
- Explain why cooperation is important in technological development.
- Identify tasks of each major area of technology.
- Explain how technological systems are related.

Answers to Test Your Knowledge *Questions, Text Pages 54–55*

1. Level of development, economic structure, number of people involved, and the type of technology developed and used
2. Evaluate individually.
3. True
4. Research and development, production, and marketing
5. Financial affairs
6. Evaluate individually.
7. False
8. People are united and combined into one body. They share a common vision, work toward similar goals, and share in the success of the enterprise.
9. f. Medical
10. c. Construction
11. e. Manufacturing
12. a. Agricultural and related biotechnologies
13. g. Transportation
14. b. Communication and information
15. d. Energy and power
16. a. Agricultural and related biotechnologies
17. b. Communication and information
18. f. Medical
19. b. Communication and information
20. Evaluate individually.

Answers to the Chapter Quiz, Teacher's Resource

1. False
2. False
3. True
4. True
5. False
6. True
7. True
8. False
9. True
10. d. Marketability of the outputs
11. a. Mylar
12. a. Research and development, production, and marketing
13. a. Communication
14. Evaluate individually.
15. Evaluate individually.

Caption for Reproducible Master

Reproducible Master 3-1. Types of Technological Systems—Shown are seven types of technology systems or contexts.

Caption for Transparency Master (Binder/CD Only)

Transparency Master 3-1. Technology Is Constantly Changing—Shown are examples of airborne technology that is obsolete, current, and emerging.

Types of Technological Systems

Name ______________________________ Period ____________

Date ______________________________ Score ____________

Chapter 3 Quiz

True/False: If a statement is true, write *True* in the blank. If a statement is false, write *False* in the blank.

_________ 1. Most technology is developed by individuals for their own use.

_________ 2. Many of yesterday's current technologies are now high technologies.

_________ 3. Construction transforms materials and manufactured goods into structures.

_________ 4. Obsolete technology is seldom used to mass-produce products.

_________ 5. Marketing involves developing the characteristics of a product or service.

_________ 6. All technology systems are interrelated.

_________ 7. Corporate participation is defined as a spirit of cooperation in developing and operating technological systems.

_________ 8. The robotic paint-spraying machines in an automobile assembly plant are one example of transportation technologies.

_________ 9. The government may develop technologies even if they are considered too expensive or too risky for private enterprise.

Multiple Choice: Choose the best answer and write the corresponding letter in the blank.

_____ 10. Which of the following is *not* one of the four general ways technology systems can be categorized?
- a. Level of development
- b. Economic structure
- c. Number of people involved
- d. Marketability of the outputs

_____ 11. Which of the following is an example of government-sponsored technological innovation?
- a. Mylar
- b. Electric lightbulb
- c. Telegraph
- d. Velcro

_____ 12. Technology moves through three stages as it meets human needs and wants. Which of the following is the proper order of these stages?
- a. Research and development, production, and marketing
- b. Production, research and development, and marketing
- c. Research and development, marketing, and production
- d. Marketing, research and development, and production

(Continued)

Name ____

_____ 13. Which of the following types of technology converts information into media messages?
a. Communication
b. Construction
c. Manufacturing
d. Transportation

Short Answer: In the space provided, write a brief answer to each of the following questions.

14. List the major types of technology systems and give examples of how you use them in your life.

____

____

____

____

____

15. Explain how technological systems are related.

____

____

____

____

____

Types of Technological Systems

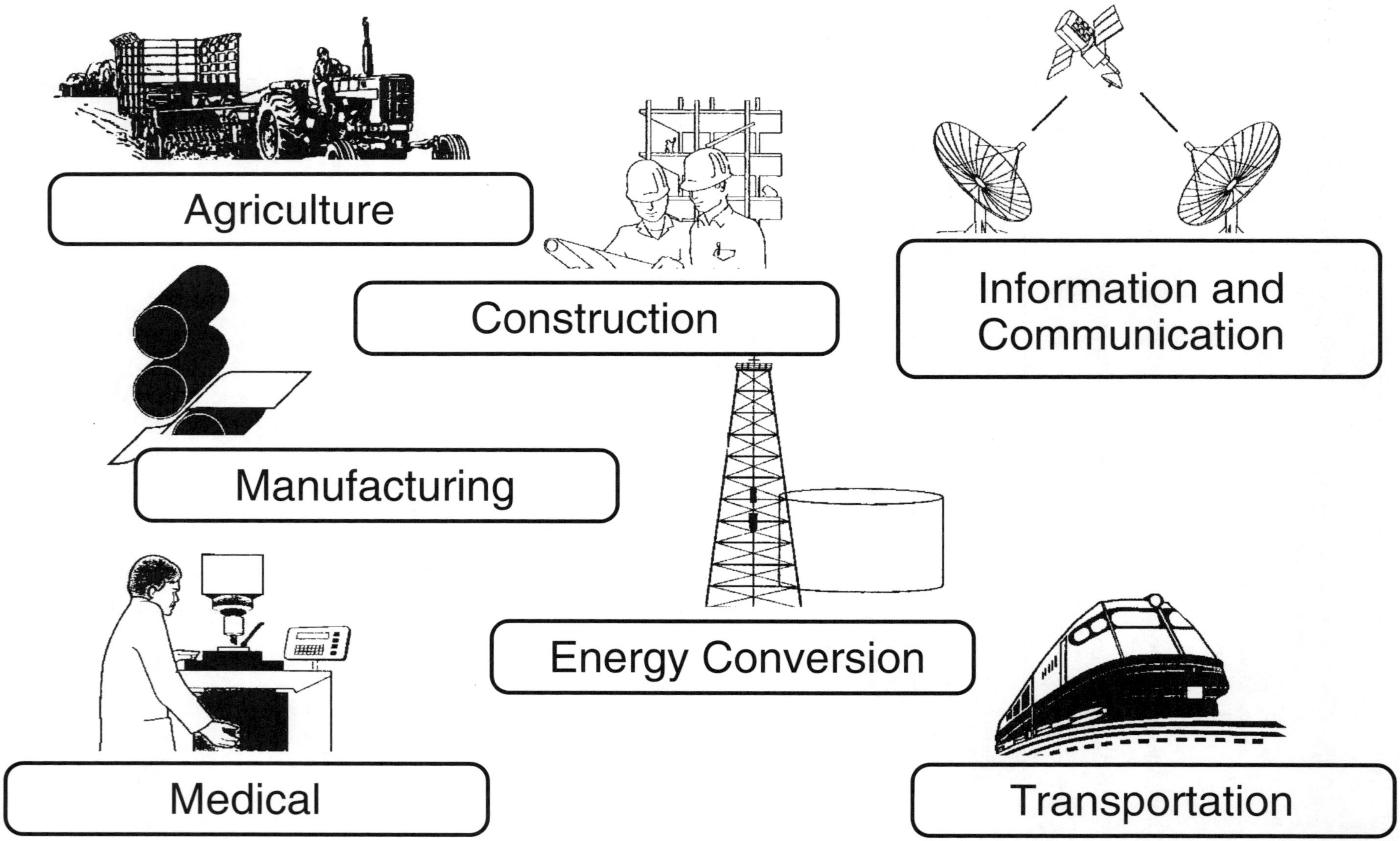

Section 2

Technological System Components

Teaching Materials

Text

Chapter 4, pp. 60–78
- Test Your Knowledge, p. 78
- Applying Your Knowledge, p. 78

Chapter 5, pp. 80–102
- Test Your Knowledge, p. 101
- Applying Your Knowledge, p. 102

Chapter 6, pp. 104–121
- Test Your Knowledge, pp. 120–121
- Applying Your Knowledge, p. 121

Section Activities
- Activity 2A—Design Problem, p. 122
- Activity 2B—Fabrication Problem, pp. 122–123

Student Activity Manual

Activity 2A—Design Problem, pp. 13–14
Activity 2B—Fabrication Problem, pp. 15–16

Teacher's Resources

Chapter 4 Quiz, pp. 81–82
Chapter 5 Quiz, pp. 88–89
Chapter 6 Quiz, pp. 94–95
Reproducible Masters
- 4-1: Basic Mechanisms, p. 83
- 4-2: Classification of Materials, p. 84
- 4-3: Solid Materials, p. 85
- 4-4: Properties of Materials, p. 86
- 5-1: Problem-Solving/Design Processes, p. 90
- 5-2: Functions of Management, p. 91
- 6-1: Types of Results from Technological Systems, p. 96

Transparencies (Binder/CD only)
- 4-1: Inputs to Technological Systems
- 4-2: Types of Tools
- 5-1: Technological Processes

Organization of Section 2

Section 2 is designed to allow students to explore the important components of all systems: inputs, processes, outputs, and feedback and control. First, students will explore the inputs common to all technology systems, Figure 2-1. These inputs are:

- People
- Tools and machines

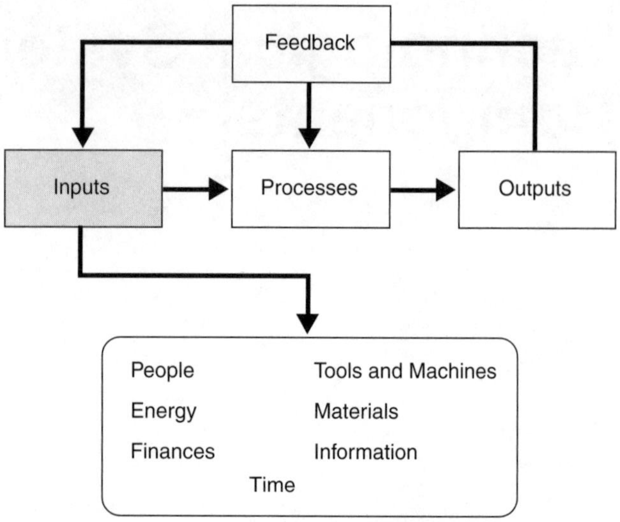

Figure 2-1

- Materials
- Information
- Energy
- Finances
- Time

Next, students will investigate the three major types of processes used in technological systems in further depth than they did in Section 1, Figure 2-2. These processes are:

- *Problem-solving/design processes*: Used to design and engineer the outputs of systems and the systems themselves
- *Production processes*: Used to grow and process crops, produce products, convert and use energy, structures, produce media messages, treat illnesses, and transport people and cargo
- *Management processes*: Used to plan, organize, actuate, and control technological systems

Finally, students will explore the outputs of technological systems and feedback and control systems, Figure 2-3.

For this section, the students spend at least five days covering the content of the three chapters and in completing a laboratory activity. These chapters are Chapter 4—Inputs to Technological Systems, Chapter 5—Technological Processes, and Chapter 6—Outputs and Feedback and Control.

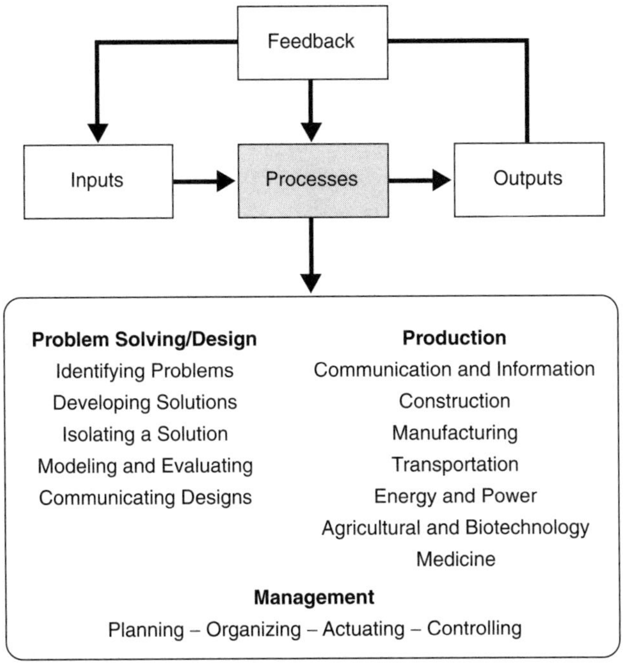

Figure 2-2

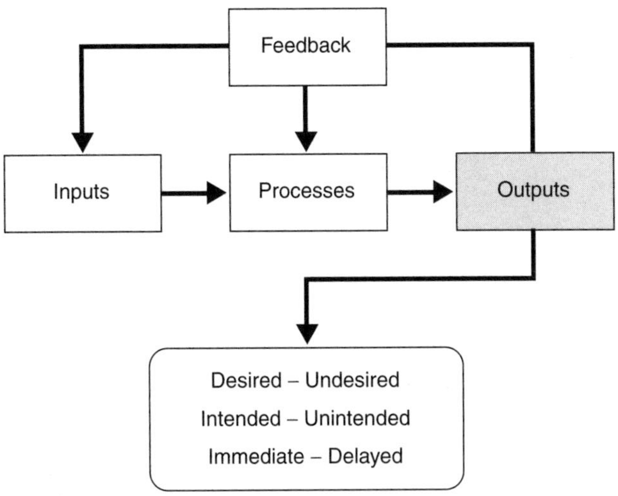

Figure 2-3

This section, like all others in this course, has two different types of activities provided. The design problem allows students to engage in a creative, problem-solving approach, whereas the fabrication problem provides a more structured approach to the laboratory experience.

Presenting Section 2

Estimated length—1 week

Section 2 helps students understand technology as a system that:

- Has materials, tools and machines, people, information, energy, finances, and time as resources.
- Uses problem-solving/design, production, and management processes.
- Has desired and undesired outputs, intended and unintended outputs, and immediate and delayed outputs.
- Uses feedback to control the outputs.

The content of Section 2 can be presented by using just lecture/discussion strategies. Note, however, that technology education is based on the concept that presentations should be reinforced with realistic laboratory activities. Thus, about 75 percent of the time assigned to this section is dedicated to laboratory activity. The five-day schedule developed for this section reflects this emphasis, Figure 2-4.

Three important topics are introduced through presentations and discussions. These are (1) inputs to technology, (2) processes of technology, and (3) outputs and feedback and controls of technology. The students will be using simple tools to complete either a design or

Section 2: Technological Systems	
Day	Activity
1	Discussion: "Inputs to Technological Systems" (Chapter 4) Introduce laboratory activity
2	Students work on laboratory activity
3	Discussion: "Technological Processes" (Chapter 5) Students continue work on laboratory activity
4	Discussion: "Outputs and Feedback and Control of Technological Systems" (Chapter 6) Complete construction part of laboratory activity
5	Students demonstrate results of laboratory activity

Figure 2-4

a fabrication problem. It is important that the students use tools and machines SAFELY. Demonstrate appropriate safety procedures prior to any laboratory activity.

To prepare to present this section, you should:

1. Read Chapters 4, 5, and 6 in *Technology.*
2. Prepare discussions on:
   a. Inputs to technology systems (Chapter 4).
   b. Processes of technology (Chapter 5).
   c. Outputs and feedback and control of technological systems (Chapter 6).
3. Select a laboratory activity.
   a. Activity 2A—This activity is a creative, problem-solving activity.
   b. Activity 2B—This activity is a structured activity. Groups of students construct a simple technological system to lift a weight. The students then analyze the system in terms of inputs, processes, and outputs.
4. Practice the selected laboratory activity.
5. Prepare teaching materials (transparencies or bulletin boards, for example) to support the content.
6. Determine which study questions and quiz questions will be used.

Presenting the Activities

Consider the following recommendations in presenting the activity you select for this section.

Activity 2A—Design Problem

In this activity, the students are challenged to design a kit that contains common materials, such as tongue depressors, mousetraps, rubber bands, thread spools, small wooden blocks, and string.

They are then asked to develop a technological device using these materials. The device should contain a power source, and it should employ at least three machines to lift a tennis ball six inches or more off a table.

To present this activity:

1. Introduce the background, situation, and challenge of the activity.
2. Divide the class into groups of two to four students. **Note:** Students may work independently on this activity.
3. Have each member of each group:
   a. Work independently to develop at least two solutions to the problem. These solutions should include (1) a supply list for the kit and (2) sketches of devices that can be built from the supplies.
   b. Present his/her solution to their group.
5. Have each group:
   a. Discuss each solution.
   b. Select a "best" solution.
   c. Develop a supply list for the kit.
   d. Construct at least one working model of the technological device using the supplies listed.
   e. Present their solution to the class.

Activity 2B—Fabrication Problem

To present this activity:

1. Discuss the background and challenge for this activity, which is included in the Section Activity in the textbook. The challenge for the students is to construct the device shown in the activity and then test it to determine its ability as a force multiplier.
2. Divide the class into groups of two or three students.
3. Have the groups follow the procedures listed on the activity sheet to build the technological device to lift a load.

Reinforcing and Enriching

The following activities can be used to supplement and enrich the laboratory activities at the end of each textbook section and the Applying Your Knowledge activities included at the end of each textbook chapter.

1. Have the students use a block of wood and a tongue depressor (craft stick) to make a lever. Have them experiment with first-class, second-class, and third-class levers. They should explain the levers as

force or distance multipliers. The students should then give examples of where each type of lever could be used.

2. Prepare a bulletin board or three-dimensional display that shows a simple product as a technological system with inputs, processes, and outputs.

Adjustments for Exceptional Students

Both laboratory activities are designed to be done in a short period of time. The design problem is open ended to accommodate varying abilities and levels of sophistication. The fabrication problem is fairly simple, and its directions can be followed easily. In a class that has a wide range of abilities, students needing more direction and structure should be assigned the fabrication problem, and gifted students should complete the design problem. When this approach is used, both groups should report their results to the entire class.

Evaluation Procedures

The understanding that students have of technology and technological systems should be measured in several ways, including:

- Observation and critical listening during class presentations and discussions.
- Worksheets and quizzes covering the cognitive material presented.
- Open-ended questions about the learned material and how technology is developed.
- Assessment of individual participation and contributions to group work.
- Assessment of the quality and creativeness of the solutions to the assigned design/fabrication activities.

Chapter 4
Inputs to Technological Systems

Objectives

After studying Chapter 4, the students will be able to:

- Identify inputs to technological systems.
- Describe the types of skills and knowledge various groups of people bring to technological systems.
- Identify the types of tools and machines used as inputs to technological systems.
- Describe the types and properties of materials that are inputs to technological systems.
- Identify the types of information that are inputs to technological systems.
- Identify the major types and sources of energy as inputs in technological systems.
- List the sources of finances that are used as inputs to technological systems.
- Explain the importance of time with regard to technological systems.
- Explain why we need to use and conserve intelligently the resources that are the inputs to technological systems.

Answers to Test Your Knowledge *Questions, Text Page 78*

1. False
2. Their attitudes about styling, price, and service directly affect the systems' outputs. Their money is spent on the products or services.
3. False
4. Pulley
5. In first-class levers, the fulcrum is between the load and the effort. In second-class levers, the load is between the effort and the fulcrum. In third-class levers, the effort is placed between the load and the fulcrum.
6. Wedge
7. Evaluate individually.
8. Liquids
9. True
10. Scientific
11. True
12. Evaluate individually.
13. Debt financing
14. False
15. The quality and quantity of inputs often determine the type of technology a society can use. Many of these inputs are in limited supply. Therefore, everyone designing, building, or using technology should use inputs appropriately and recycle as many of them as possible.

Answers to the Chapter Quiz, Teacher's Resource

1. False
2. True
3. False
4. False
5. True
6. False
7. True
8. False
9. True
10. c. Engineers
11. a. The wedge and screw
12. c. Genetic material
13. c. Gases
14. b. Acoustical properties
15. The seven inputs are people, tools and machines, materials, information, energy, finances, and time. Evaluate examples individually.

Captions for Reproducible Masters

Reproducible Master 4-1. Basic Mechanisms—Shown are examples of the lever principle and the inclined plane principle. Each of the six simple mechanisms comes from one of these two principles.

Reproducible Master 4-2. Classifications of Materials—Shown are four different ways in which materials can be classified. Each classification tells something useful about the material.

Reproducible Master 4-3. Solid Materials—Shown are the four categories of solid materials.

Reproducible Master 4-4. Properties of Materials—Shown are the seven categories of material properties. All materials can be described by their characteristics in these seven areas.

Captions for Transparency Masters (Binder/CD Only)

Transparency Master 4-1. Inputs to Technological Systems—Shown are the seven categories of inputs to technological systems.

Transparency Master 4-2. Types of Tools—Tools can be grouped into a number of categories. Shown are six common categories of tools with examples for each group.

Inputs to Technological Systems

Name ____ Period ____________________

Date ____ Score _____________________

Chapter 4 Quiz

True/False: If a statement is true, write *True* in the blank. If a statement is false, write *False* in the blank.

_________ 1. Engineers develop the basic knowledge needed to develop products and processes.

_________ 2. Technicians are skilled workers and usually work in laboratories or product-testing facilities.

_________ 3. Drilling tools are used to separate materials into two or more pieces.

_________ 4. The fulcrum is close to the force when a force multiplier lever is used.

_________ 5. Solids can support loads without losing their shape.

_________ 6. Scientific information concerns the design and production of human-made products.

_________ 7. Thermal energy is created by increased molecular action caused by heat.

_________ 8. Raising money by selling a portion of the company is called debt financing.

_________ 9. The technology that a society can use is determined by the quality and quantity of inputs.

Multiple Choice: Choose the best answer and write the corresponding letter in the blank.

_____ 10. Which group of people determines the appropriate materials and processes needed to produce products or perform services?
a. Scientists
b. Technicians
c. Engineers
d. Production workers

_____ 11. Which of the following operate on the principle of the inclined plane?
a. The wedge and screw
b. The inclined plane and pulley
c. The wheel and axle and inclined plane
d. The lever and wedge

_____ 12. Wood can be classified as a(n):
a. inorganic material.
b. polymeric material.
c. genetic material.
d. exhaustible material.

_____ 13. Which of the following types of material can be easily compressed?
a. Solids
b. Liquids
c. Gases
d. Composites

(Continued)

Name ____

_____ 14. The reaction of a material to sound waves is called its:
a. thermal properties.
b. acoustical properties.
c. electrical and magnetic properties.
d. optical properties.

Short Answer: In the space provided, write a brief answer to the following question.

15. List and give an example of each of the seven inputs to technology systems.

____

____

____

____

____

____

____

Basic Mechanisms

Lever Principle
- Lever
- Wheel and Axle
- Pulley

Inclined Plane Principle
- Wedge
- Inclined Plane
- Screw

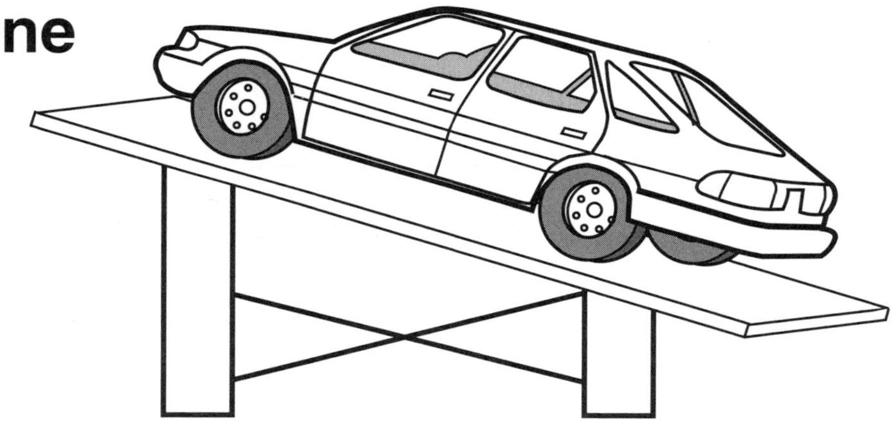

Classification of Materials

Natural, Synthetic, or Composite

Exhaustible or Genetic

Organic or Inorganic

Gas, Liquid, or Solid

Solid Materials

Materials that hold their size and shape.

Ceramic

Metallic

Polymeric

Composite

Properties of Materials

Physical

Mechanical

Chemical

Thermal

Electrical and Magnetic

Acoustical

Optical

Chapter 5
Technological Processes

Objectives

After studying Chapter 5, the students will be able to:

- Describe the three major types of technological processes and their relationship to one another.
- Identify the steps used in the problem-solving/design process.
- Explain the major production steps used in the agricultural and related biotechnological, communication and information, construction, energy and power, manufacturing, medical, and transportation processes.
- Identify the steps involved in management processes.

Answers to Test Your Knowledge *Questions, Text Page 101*

1. Management
2. False
3. Evaluate individually.
4. False
5. Conversion/processing
6. True
7. Evaluate individually.
8. Foundation
9. c. Mechanical
10. Harvesting, mining, and drilling
11. True
12. Treatment
13. Land
14. Structure, propulsion, guidance, control, and suspension
15. Planning, organizing, actuating, and controlling

Answers to the Chapter Quiz, Teacher's Resource

1. True
2. True
3. False
4. False
5. True
6. False
7. False
8. True
9. True
10. a. Apartments
11. c. Forming
12. d. Planning, organizing, actuating, and controlling
13. a. Terminals
14. d. Guidance
15. The process of transforming and applying energy involves gathering or collecting the resources, converting the energy into a new form, transmitting the energy, and applying the energy to do work. Evaluate the examples individually.

Captions for Reproducible Masters

Reproducible Master 5-1. Problem-Solving/Design and Management Processes—The same processes are used to solve problems in all seven technology clusters.

Reproducible Master 5-2. Functions of Management—Shown are the four functions of management. These functions are used to guide and direct other processes.

Caption for Transparency Master (Binder/CD Only)

Transparency Master 5-1. Technological Processes—Shown are the three important kinds of processes. These processes are used to develop and operate technological systems.

Technological Processes

Name ______________________________ Period ______________

Date ______________________________ Score ______________

Chapter 5 Quiz

True/False: If a statement is true, write *True* in the blank. If a statement is false, write *False* in the blank.

_________ 1. The first step in the problem-solving/design process includes listing limitations that communicate expectations for the solution.

_________ 2. Graphic models are often developed on a computer before physical models are constructed.

_________ 3. The first step in the communication process is to decode the message.

_________ 4. Secondary processes convert resources into industrial materials.

_________ 5. The final step in the problem-solving/design process includes specifying materials to be used in production.

_________ 6. Each technological system uses the same type of production process.

_________ 7. Transportation support systems are the onboard technical systems that make a vehicle work.

_________ 8. Although there are several types of production processes, the same design process can be used for all of them.

_________ 9. Telecommunication processes depend on electromagnetic waves to carry their messages.

Multiple Choice: Choose the best answer and write the corresponding letter in the blank.

_____ 10. Which of the following is an example of a residential structure?
a. Apartments
b. Power plants
c. Communication towers
d. Communication studios

_____ 11. Using force applied to a die to reshape material is called:
a. separating.
b. casting.
c. forming.
d. conditioning.

_____ 12. Management involves four steps. The proper order is:
a. actuating, planning, organizing, and controlling.
b. planning, actuating, organizing, and controlling.
c. organizing, planning, actuating, and controlling.
d. planning, organizing, actuating, and controlling.

(Continued)

Name ____

_____ 13. Which of the following is *not* a type of vehicle system?
a. Terminals
b. Structure
c. Propulsion
d. Control

_____ 14. Which of the following gathers and displays information so a vehicle can be kept on course?
a. Propulsion
b. Control
c. Suspension
d. Guidance

Short Answer: In the space provided, write a brief answer to the following question.

15. Describe the process of transforming and applying energy and give a specific example.

____

____

____

____

____

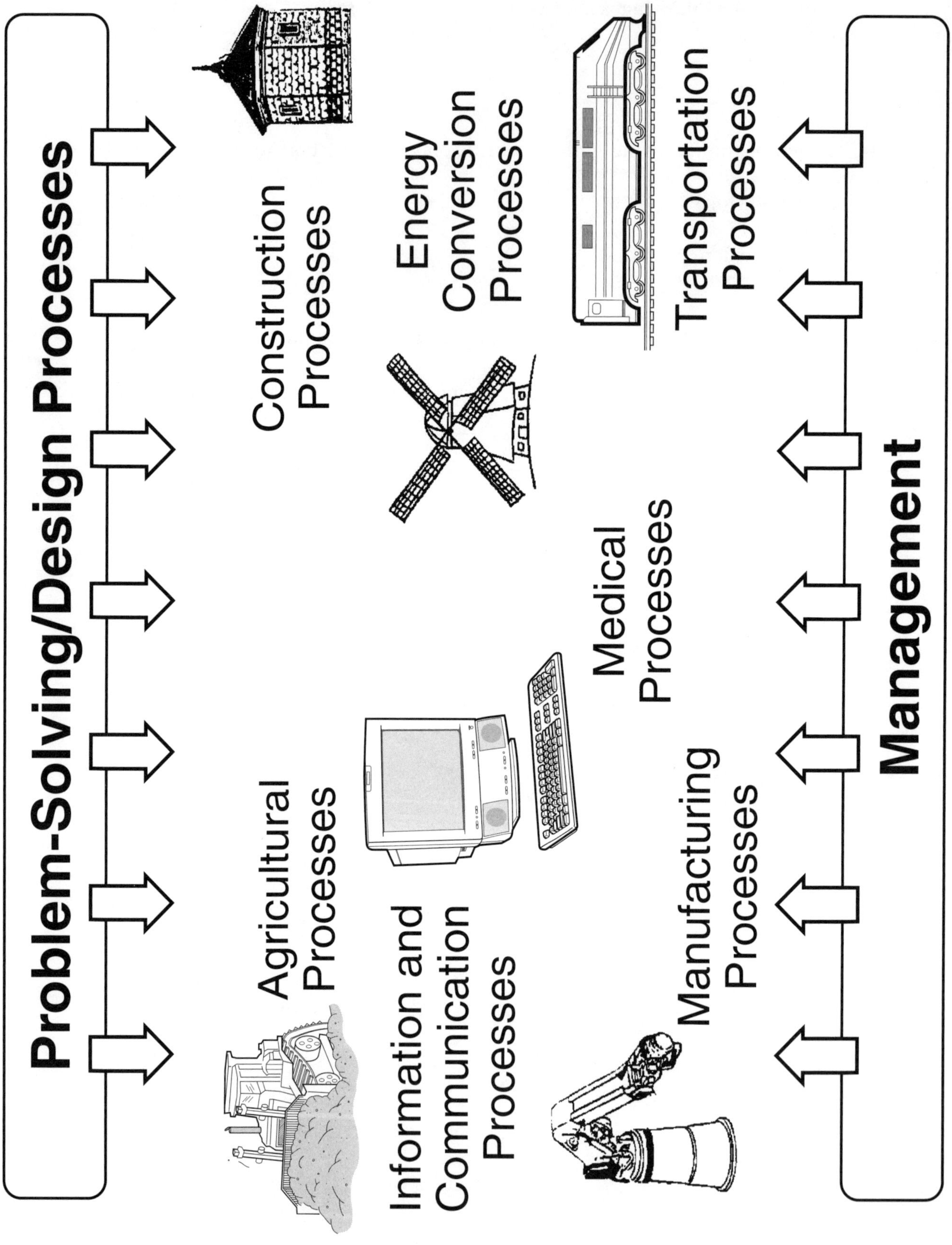
Problem-Solving/Design Processes
Agricultural Processes
Construction Processes
Information and Communication Processes
Energy Conversion Processes
Medical Processes
Manufacturing Processes
Transportation Processes
Management

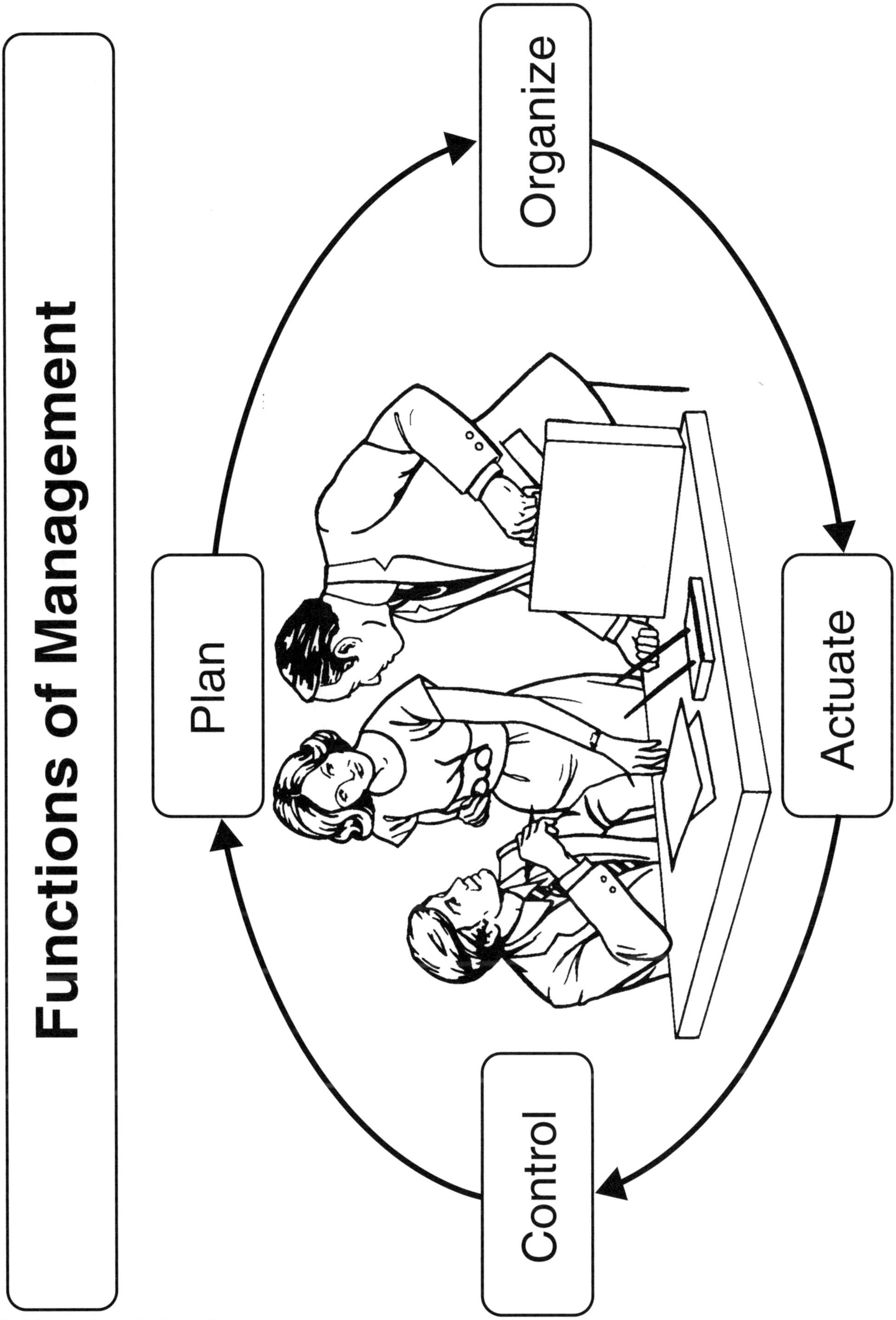
Functions of Management
Plan
Organize
Actuate
Control

Chapter 6
Outputs and Feedback and Control

Objectives

After studying Chapter 6, the students will be able to:

- Describe the relationship between outputs and feedback and control in technology systems.
- Identify the major types of outputs of various technology systems.
- Differentiate between desired and undesired, intended and unintended, and immediate and delayed outputs of technological systems.
- Define feedback as used in technology systems.
- Differentiate between the two major types of internal control systems.
- Describe the major components of internal control systems.
- Describe the difference between manual and automatic control in internal control systems.
- Identify external controls in technology systems.

Answers to Test Your Knowledge *Questions, Text Pages 120–121*

1. Closed-loop control systems use feedback to regulate the system or its inputs to affect the output.
2. Evaluate individually.
3. True
4. Evaluate individually.
5. Output
6. Output information is used to adjust the processing actions in closed-loop control systems. Output information is not used to adjust the process in open-loop control systems.
7. c. Optical
8. b. Mechanical
9. Evaluate individually.
10. True

Answers to the Chapter Quiz, Teacher's Resource

1. False
2. False
3. True
4. False
5. True
6. False
7. True
8. True
9. c. Minimize undesirable outputs and maximize desirable outputs
10. b. Closed loop control systems
11. a. Cams
12. a. Fluidic controllers
13. d. Public opinion
14. a. Quality control
15. The monitoring device gathers information about the action being controlled. The information gathered by monitoring devices is matched against expectations by data-comparing devices. Adjusting devices modify the system to produce better outputs.

Caption for Reproducible Master

Reproducible Master 6-1. Types of Results from Technological Systems—Shown are three different categories of results from a technological system.

Outputs and Feedback and Control

Name ______________________________ Period ______________

Date ______________________________ Score ______________

Chapter 6 Quiz

True/False: If a statement is true, write *True* in the blank. If a statement is false, write *False* in the blank.

_________ 1. Intended outputs are always desirable outputs.

_________ 2. The goal of inventory control is to reduce the raw material inventory.

_________ 3. Open-loop control systems do not use feedback to control output.

_________ 4. A thermostat is an example of a manual control system.

_________ 5. Analytical systems can be used for managerial control.

_________ 6. Most technological systems use open-loop control systems.

_________ 7. Monitoring devices are used to gather information that is matched against expectations.

_________ 8. Jet pilots use both manual and automatic control systems.

Multiple Choice: Choose the best answer and write the corresponding letter in the blank.

_____ 9. Control systems are designed to:
a. maximize delayed outputs and eliminate undesirable outputs.
b. minimize desirable outputs and maximize undesirable outputs.
c. minimize undesirable outputs and maximize desirable outputs.
d. maximize all types of output.

_____ 10. Output information is used to affect output in:
a. open-loop control systems.
b. closed loop control systems.
c. Both a and b.
d. Neither a nor b.

_____ 11. Which of the following are mechanical controllers?
a. Cams
b. Electromagnetic coils
c. Switches
d. Diodes

_____ 12. Which of the following could use air to adjust machines?
a. Fluidic controllers
b. Mechanical controllers
c. Electrical controllers
d. None of the above

(Continued)

Name ____

_____ 13. Which of the following is an example of an external control?
a. A vehicle's emissions control system
b. A thermostat
c. An inventory control system
d. Public opinion

_____ 14. Which type of control ensures that acceptable outputs are produced?
a. Quality control
b. Inventory control
c. Material resource control
d. Process control

Short Answer: In the space provided, write a brief answer to the following question.

15. Describe the major components of internal control systems.

____

____

____

____

____

Types of Results from Technological Systems

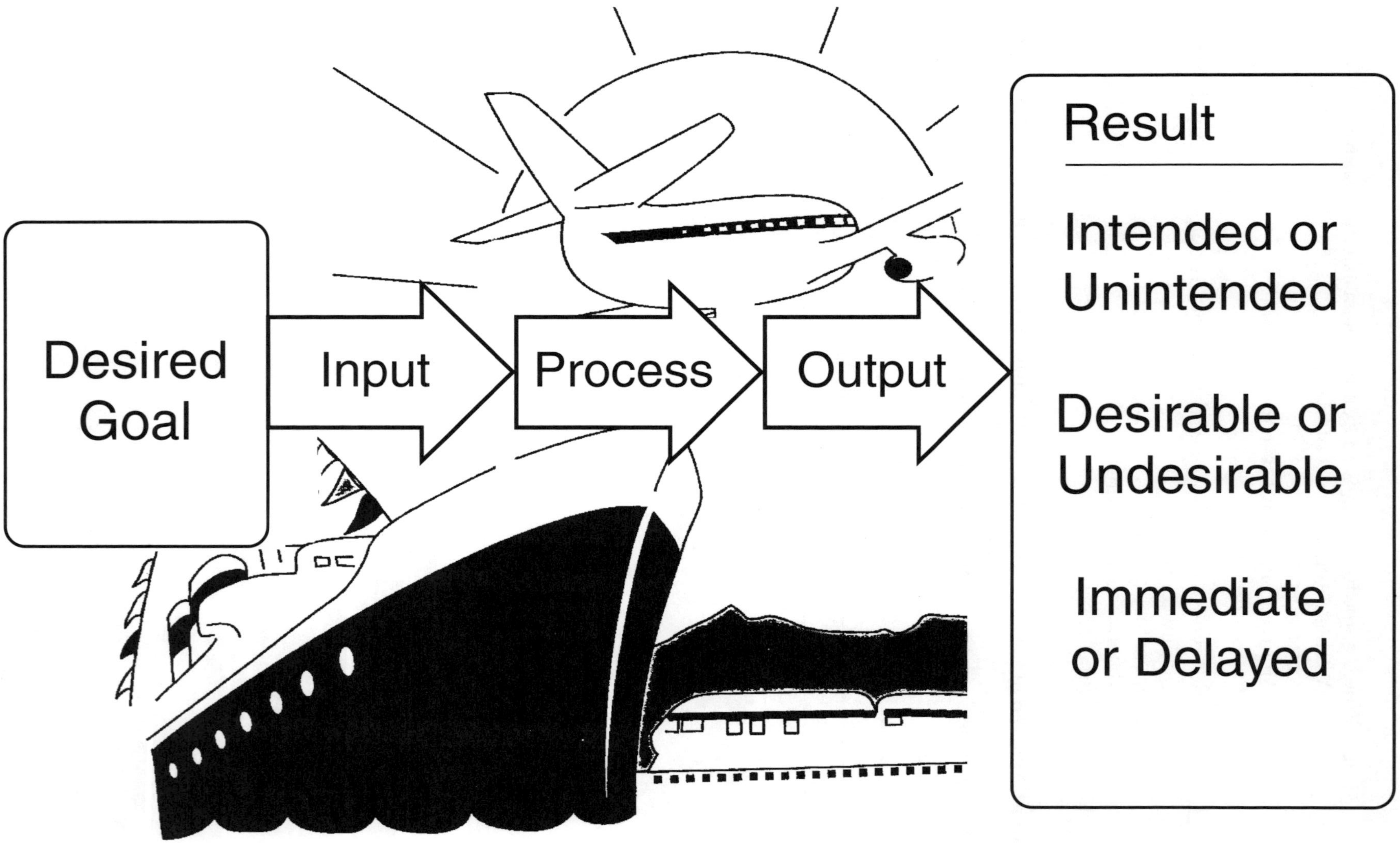

Section 3

Tools of Technology

Teaching Materials

Text

Chapter 7, pp. 126–155
    Test Your Knowledge, pp. 154–155
    Applying Your Knowledge, p. 155
Chapter 8, pp. 156–173
    Test Your Knowledge, pp. 172–173
    Applying Your Knowledge, p. 173
Section Activities
    Activity 3A—Design Problem, p. 174
    Activity 3B—Fabrication Problem, pp. 174–175

Student Activity Manual

Activity 3A—Design Problem, pp. 19–21
Activity 3B—Fabrication Problem, pp. 23–25
Taking Your Technology Knowledge Home, pp. 27–29

Teacher's Resources

Chapter 7 Quiz, pp. 105–106
Chapter 8 Quiz, pp. 115–116
Reproducible Masters
    7-1: Types of Motion, p. 107
    7-2: Sawing Machines, p. 108
    7-3: Watt Steam Machine, p. 109
    7-4: Electric Motor, p. 110
    7-5: Parts of a Computer, p. 111
    8-1: Measurement, p. 117
Transparencies (Binder/CD only)
    7-1: Types of Technological Processing
    7-2: Internal Combustion Engine
    7-3: Radio Transmission
    8-1: Common Factors That Are Measured

Organization of Section 3

Section 3 has been developed to introduce the common tools and machines of technology and to give students an opportunity to use these tools and machines. The students will be introduced to tools and machines used in three major types of processing, Figure 3-1:

- Material processing
- Energy processing
- Information processing

When the students study material-processing tools and machines, they will explore the three essential elements that make machine tools what they are, Figure 3-2:

- The cutting element or tool: A device (tool) that removes excess material to produce the desired size and shape
- Motion: The movement between the work and the tool that produces the cutting force (cutting motion) and brings new material into the cutting area (feed motion)
- Support: Devices such as tables, arbors, or chucks that support and/or rotate the work or the tool

The students will apply their knowledge of these three essential elements as they look at the major types of machine tools: turning, sawing, drilling, shaping, planing, and grinding. Following an introduction to material-processing tools and machines, the students will study three common energy-processing converters that have provided the foundation for the development of industry and advanced technology. These are the steam engine, the internal combustion engine, and the electric motor. This study will be

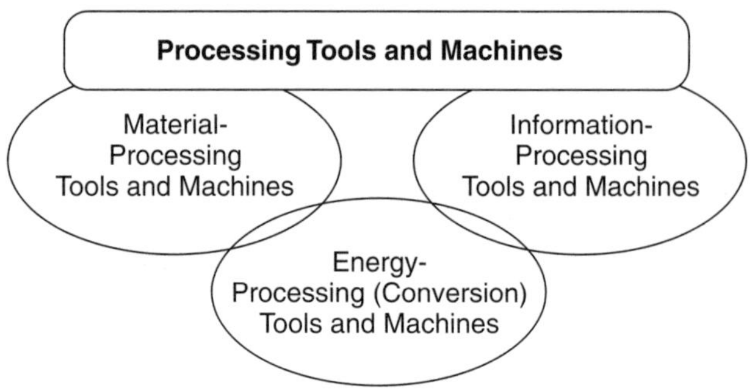

Figure 3-1

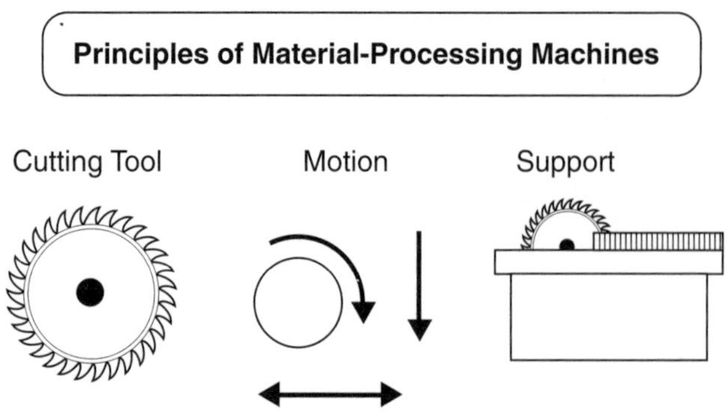

Figure 3-2

followed by an opportunity for the students to study some common information-processing machines. These include the printing press, telephone, radio, and computer. The specific focus will be on improving our ability to communicate information and ideas. The computer will be introduced as the most developed information-processing machine. It includes five specific units, Figure 3-3:

- Input unit: Devices used to enter data (numbers and letters) into the system
- Processing unit: The part of the computer, called a central processing unit (CPU) or microprocessor, that handles and manipulates the data
- Output unit: Devices used to display and record the results of the processing unit's actions
- Memory unit: The section of the computer that holds information and instructions
- Program: The directions the computer uses to process the data and to produce the output

The students will then learn the guidelines for using technological devices safely. They will learn general safety rules and those specific to working safely with people, materials, and tools and machines.

The students will then be introduced to measurement systems and tools. They will explore the two major types of measuring systems: the U.S. customary system and the International System of Units (abbreviated SI from the French name *Système International d'Unités*) or metric system. They will learn that people have developed systems to measure size and space, mass, time, temperature, number of particles, electric current, and light intensity. They will also learn about standard and precision measurement and how to apply measurement to production and quality control activities. For this section, the students will spend approximately three weeks exploring material-, energy-, and information-processing tools and machines; general and specific safety rules; and various measurement systems and tools. They will spend their time covering and applying the content of the two chapters assigned to this section. These chapters are:

- Chapter 7—Production Tools and Their Safe Use
- Chapter 8—Measurement Systems and Tools and Their Role in Technology

As with most of the sections in this program, two different types of activities are provided. The design problem lets students engage in a creative, problem-solving approach to using information-processing technology. The fabrication problem provides a structured approach to using material-processing technology. Measurement should be stressed in both activities. Also, if time permits, you can use both activities to give students an opportunity to experience the two different types of processing technology and both design and fabrication practices.

Computer Components

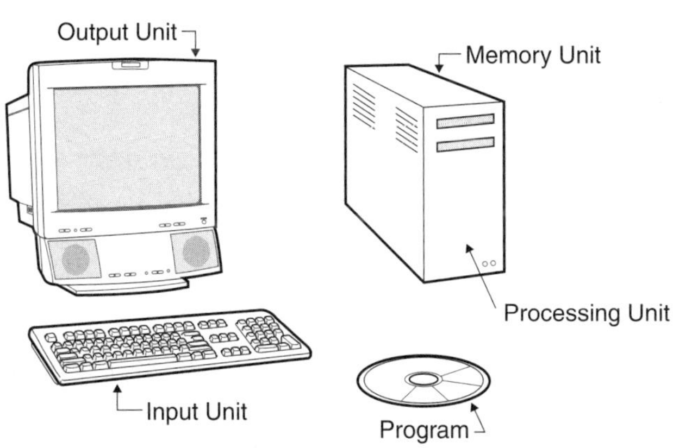

Figure 3-3

Presenting Section 3

Estimated length—3 weeks

Section 3 is designed to help the students understand that all technology processes use inputs to make more useful outputs. The basic inputs are materials, energy, and information. Each type of processing relies on measurement, as does all technological activity. Section 3 lends itself to laboratory activities. These activities include information processing, which is presented in Activity 3A—Design Problem;

material processing, which is presented in Activity 3B—Fabrication Problem; and energy processing, which can be presented in an alternate activity developed by you. Also, each process will be practiced in activities included in later sections. The 15-day schedule shown in Figure 3-4 has been developed for this section. During this period of time, you will notice that four major topics are introduced through presentations and discussions. In addition, the students will use simple processing techniques to complete either a design or fabrication problem. **Note:** It is important to explain how to use the tools and machines SAFELY. Demonstrate appropriate safety procedures prior to any laboratory activity. To prepare to present this section, you should:

1. Read Chapters 7 and 8 in *Technology.*
   a. Material-processing tools and machines (Chapter 7).
   b. Energy-converting tools and machines (Chapter 7).
   c. Information-processing tools and machines (Chapter 7). **Note:** More detailed experiences in material processing, energy processing, and information processing will be presented in Sections 5, 6, and 8.
   d. Safety rules when working with other people, materials, and tools and machines (Chapter 7).
   e. Measurement systems and tools (Chapter 8).
2. Select a laboratory activity.
   a. Activity 3A—This activity is an open-ended problem-solving activity that uses information-processing tools.
   b. Activity 3B—This activity is a structured activity where groups of students use material-processing tools to produce a simple product.
3. Practice the selected laboratory activity.
4. Prepare teaching materials (transparencies or bulletin boards, for example) to support the content.
5. Determine which study questions and quiz questions will be used.

Section 3: Tools of Technology	
Day	Activity
1	Discussion: "Tools and Technology" Introduce laboratory activity
2	Discussion: "Material-Processing Tools" (Chapter 7) Demonstrate laboratory activity procedures
3–4	Students work on laboratory activity
5	Discussion: "Energy-Converting Tools" (Chapter 7) Students work on laboratory activity
6	Discussion: "Information-Processing Tools" (Chapter 7) Students work on laboratory activity
7	Students work on laboratory activity
8	Discussion: "Tools and Safety" (Chapter 7) Students work on laboratory activity
9	Discussion: "Measurement Systems and Tools" (Chapter 8) Students work on laboratory activity
10	Students work on laboratory activity or Introduce second laboratory activity
11–14	Students complete laboratory activity
15	Students present results of activity Summarize section

Figure 3-4

Presenting the Activities

Consider the following recommendations in presenting the activity you select for this section.

Activity 3A—Design Problem

During this activity, the students will be challenged to gather information about an issue they believe is not being addressed by local politicians and business leaders, such as the greenhouse effect, fossil fuel dependence, or nuclear power. They are then to develop a one-page informational flyer to communicate (information processing) a position on the issue. If computers are available at your school, have students use one or more computer software programs to produce the flyer. To conduct this activity:

1. Divide the class into groups of three to five students.
2. Have each student in each group:
   a. List three issues that he or she thinks are important.
   b. Present his or her list to the group.
   c. Participate in discussions on each issue.
3. Have each group:
   a. Select one of the issues presented by its members.
   b. Select a theme for its presentation.
4. Have each student in each group:
   a. Develop copy for the flyer.
   b. Select or develop graphics for the flyer.
   c. Prepare a rough layout of the flyer.
5. Have each group:
   a. Select the best layout(s) for the flyer from the ones developed by its members.
   b. Brainstorm ways to integrate ideas from several layouts on how to improve the selected layout.
   c. Prepare a final layout of the flyer.
   d. Present the layout to the class.

Activity 3B—Fabrication Problem

This activity will help students understand that they live in a material world, where products all around them have been developed using material-processing technology. Each student will work with a partner to manufacture a game that can be given to a local charity. To conduct the activity:

1. Prepare or obtain the materials needed to make the product: one piece ¾" × 3½" × 3½" clear pine, redwood, or western red cedar and twelve ⅛" diameter × 1" wood pegs or golf tees.
2. Demonstrate the procedure needed to produce the product. (This is included in the operation process chart in the textbook description of this activity.)
3. Have the students:
   a. Carefully observe the demonstration and list the safety practices to be followed. Note: Stress that they should follow all the safety rules discussed during the demonstration.
   b. Construct the game following the procedure demonstrated in class and presented in the operation process chart.
4. Complete the analysis exercise for this activity.

Reinforcing and Enriching

The following activities can be used to supplement and enrich the laboratory activities at the end of each textbook section and the Applying Your Knowledge activities included at the end of each textbook chapter.

1. Have the students select an electrically powered product and analyze it in terms of the:
   a. Material-processing tools used to make it.
   b. Energy conversion technology it uses.
   c. Information-processing activities used to produce the package.
2. Have the students who completed Activity 3B—Fabrication Problem develop a package for the product using information-processing practices.

Taking Your Technology Knowledge Home

This take-home activity allows students to apply their knowledge of tools to design a special tool for work around their home. They will:

1. Identify a task for which they could design a tool.
2. Research similar tools in magazines and catalogs.

3. List the features the researched tools have.
4. List the features they will include in their tool.
5. Sketch four possible designs for the tool.
6. Select and refine the best design.
7. Build a prototype of the tool.
8. Ask one or more people to try out the tool.
9. List suggestions for improving the tool the users made.
10. Prepare a sketch for an improved design for the tool.

Adjustments for Exceptional Students

Both laboratory activities are designed to accommodate students of all ability levels. Activity 3A is good for gifted students who can develop a wide range of creative answers and produce more complex solutions. Lower achievers can still solve the design problem but with solutions within their ability level.

Activity 3B is simple enough to accommodate students of several dexterity levels. You can build the tooling to allow less-skilled students to produce an acceptable product.

In all cases, the amount of structure and teacher guidance should be adjusted to meet the students' needs. Students should be given the minimum amount of structure they need to be successful. Practices that hamper creativity should be avoided.

Careful assignment of groups will allow students to contribute their own special skills and abilities to the activity. Using student teams also allows students to develop cooperative work habits.

Evaluation Procedures

The understanding students have of technology and technological systems should be measured in several ways, including:

- Observation and critical listening during class presentations and discussions.
- Worksheets and quizzes covering the cognitive material presented.
- Open-ended questions about the learned material and how technology is developed.
- Assessment of individual participation and contributions to group work.
- Assessment of the quality and creativeness of the solutions to the assigned design and fabrication problems.

Chapter 7
Production Tools and Their Safe Use

Objectives

After studying Chapter 7, the students will be able to:

- List the categories of production tools used in technology.
- List the characteristics of machine tools.
- Identify the major types of machine tools.
- Describe the major types of energy-processing tools and machines.
- Describe the major types of information-processing tools and machines.
- List guidelines for using tools and machines properly and safely.

Answers to Test Your Knowledge *Questions, Text Pages 154–155*

1. Evaluate individually.
2. Cutting motion is the action that causes material to be removed from the work. It causes the excess material to be cut away. Feed motion is the action that brings new material into the cutter. It allows the cutting action to be continuous.
3. True
4. Scroll
5. Counterboring is producing two holes around the same center point. The larger diameter hole is drilled to a lesser depth than the smaller diameter hole.
6. The metal shaper moves the tool back and forth over the workpiece to produce the cutting motion. The metal planer reciprocates the workpiece under the tool to generate the cutting motion.
7. Steam engine
8. An electric motor
9. Radio waves
10. Input unit, processing unit, output unit, memory unit, and program
11. True
12. Evaluate individually.
13. Student answers will vary but may include any of the following: avoid horseplay; do not talk to anyone who is using a machine, except when necessary; do not cause other people harm by carelessly leaving tools on benches or machines. Additional responses may be accepted at the instructor's discretion.
14. Student answers will vary but may include any of the following: be careful when moving long pieces of material; use extreme caution when handling sheet metal; grip hot materials with pliers or tongs; wear gloves when handling hot or sharp materials; use extreme caution in handling hot liquids and molten metals. Additional responses may be accepted at the instructor's discretion.
15. False

Answers to the Chapter Quiz, Teacher's Resource

1. True
2. False
3. False
4. True
5. False
6. False
7. True
8. True
9. c. Reciprocating motion
10. d. Knurling
11. b. Band saws, circular saws, and scroll saws
12. a. Counterboring
13. c. Compression
14. b. Working alone whenever possible to prevent injury to others
15. Material processing, energy processing, and information processing. Student examples will vary. Evaluate individually.

Captions for Reproducible Masters

Reproducible Master 7-1. Types of Motion—Shown are the two important types of motion between a cutting tool and the workpiece. One type is the motion of the cutting tool. The other type is the motion in which the workpiece is fed into the blade.

Reproducible Master 7-2. Sawing Machines—Shown are three types of sawing machines. Each saw makes use of a different combination of cutting and feed motions.

Reproducible Master 7-3. Watt Steam Engine—Shown are the forward and backward motions of the piston in a Watt double-action steam engine.

Reproducible Master 7-4. Electric Motor—Electric motors operate using a stationary magnet and an armature that switches its magnetic poles.

Reproducible Master 7-5. Parts of a Computer—Shown are the five parts that make up a computer system.

Captions for Transparency Masters (Binder/CD Only)

Transparency Master 7-1. Types of Technological Processing—Shown are three types of technological processing. Each type of processing has its own selected set of tools.

Transparency Master 7-2. Internal Combustion Engine—Shown are the four strokes of the four-stroke internal combustion engine.

Transparency Master 7-3. Radio Transmission—This diagram shows a sound signal traveling from a microphone to a radio speaker.

Production Tools and Their Safe Use

Name ______________________________ Period ______________

Date ______________________________ Score ______________

Chapter 7 Quiz

True/False: If a statement is true, write *True* in the blank. If a statement is false, write *False* in the blank.

_________ 1. Abrasive paper is a multiple-point cutting tool.

_________ 2. The tailstock center of a lathe provides the force to rotate the workpiece.

_________ 3. Straight turning produces a cylindrical shape with a uniformly decreasing diameter.

_________ 4. Scroll saws use a reciprocating cutting motion.

_________ 5. Shapers, planers, and broaches are all single-point tools.

_________ 6. A two-stroke cycle engine moves a piston up and back twice to produce a power stroke.

_________ 7. Electric motors are the most commonly used source of power.

_________ 8. Unlike computers, calculators have fixed programs that cannot be changed.

Multiple Choice: Choose the best answer and write the corresponding letter in the blank.

_____ 9. Which of the following moves a cutting tool back and forth?
a. Rotating motion
b. Linear motion
c. Reciprocating motion
d. None of the above

_____ 10. Which of the following produces a surface that is easy to grip and turn?
a. Facing
b. Grooming
c. Chamfering
d. Knurling

_____ 11. The three basic types of saws are:
a. circular saws, band saws, and table saws.
b. band saws, circular saws, and scroll saws.
c. scroll saws, band saws, and radial saws.
d. None of the above.

_____ 12. Producing two holes of different depth and diameter around the same center point is called:
a. counterboring.
b. countersinking.
c. drilling.
d. reaming.

(Continued)

Name ____

_____ 13. Combustion takes place at the end of which stroke?
a. Intake
b. Power
c. Compression
d. None of the above

_____ 14. It is always important to follow safety guidelines when using technological devices. These guidelines include all of the following, *except:*
a. reading owner's manuals or instructions before operating any device.
b. working alone whenever possible to prevent injury to others.
c. choosing the correct tool for the job.
d. using devices for only their intended applications.

Short Answer: In the space provided, write a brief answer to the following question.

15. List and give an example of each of the three types of production tools that are used in technology.

____

____

____

Types of Motion between a Cutting Tool and a Workpiece

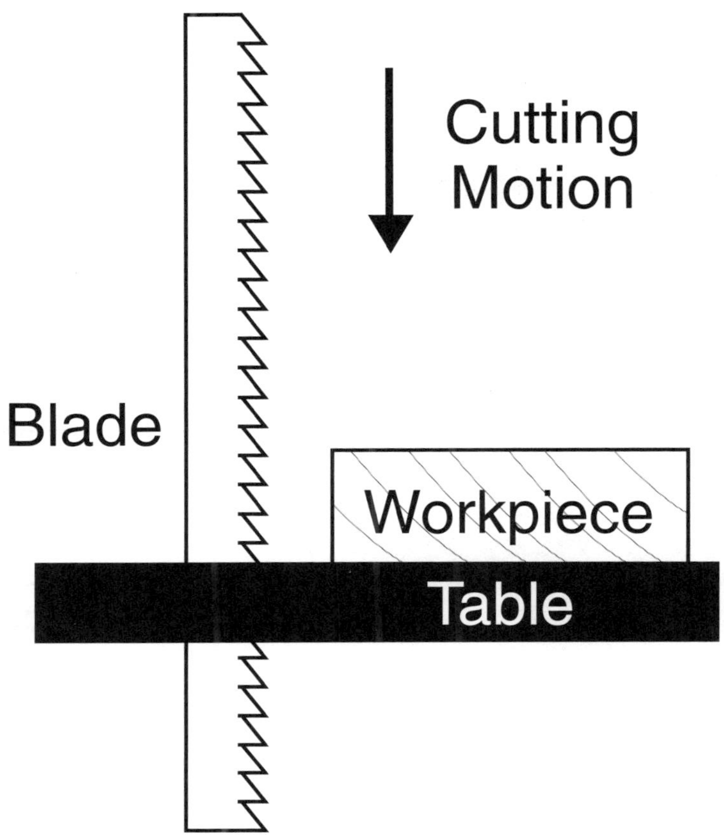

Cutting Motion

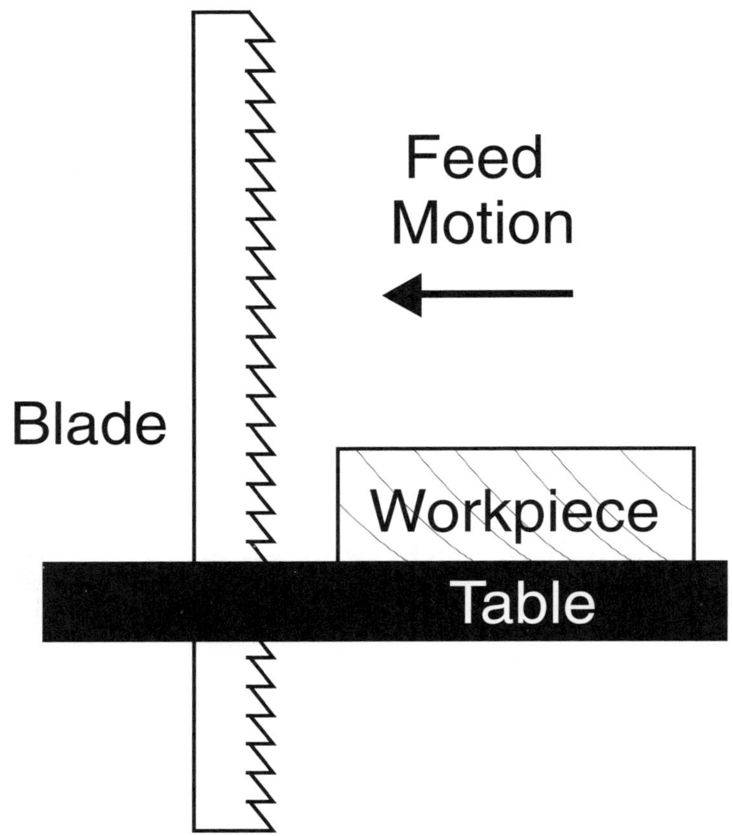

Feed Motion

Sawing Machines

Circular Saw	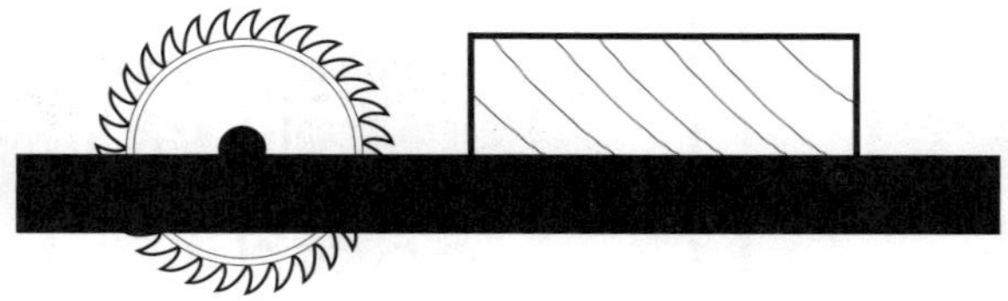	Cutting Motion-Rotating Tool Feed Motion-Linear Work Movement
Band Saw	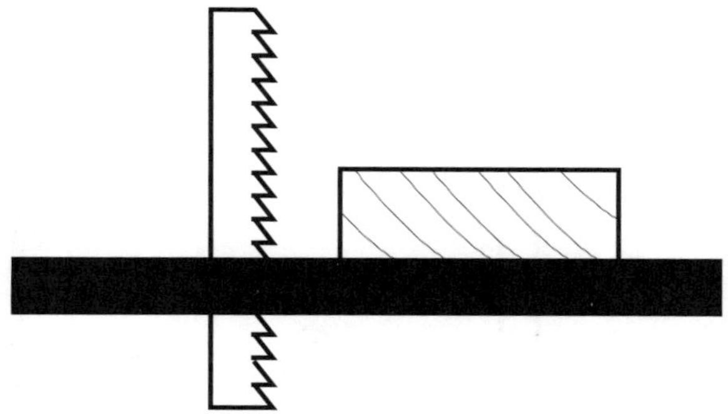	Cutting Motion-Linear Tool Movement Feed Motion-Linear Work Movement
Scroll Saw	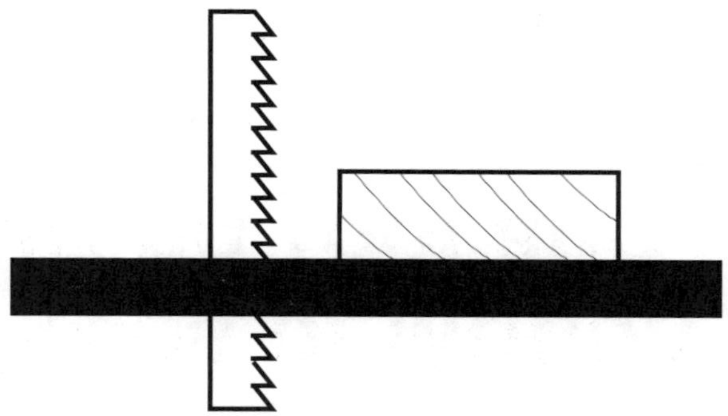	Cutting Motion-Reciprocating Tool Feed Motion-Linear Work Movement

Watt Steam Machine

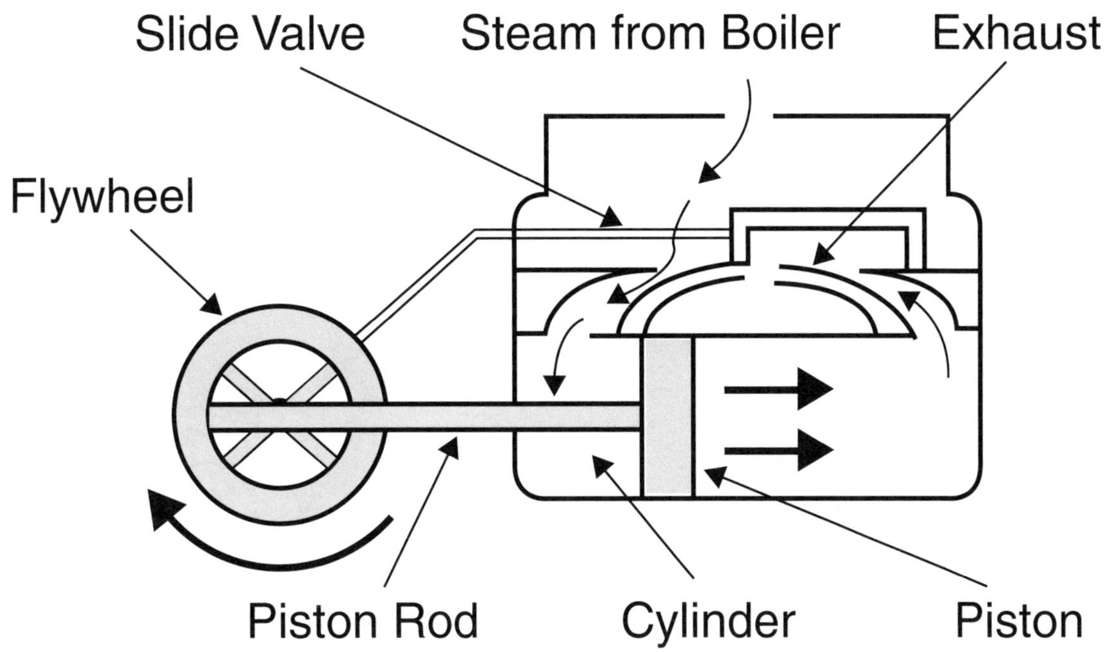

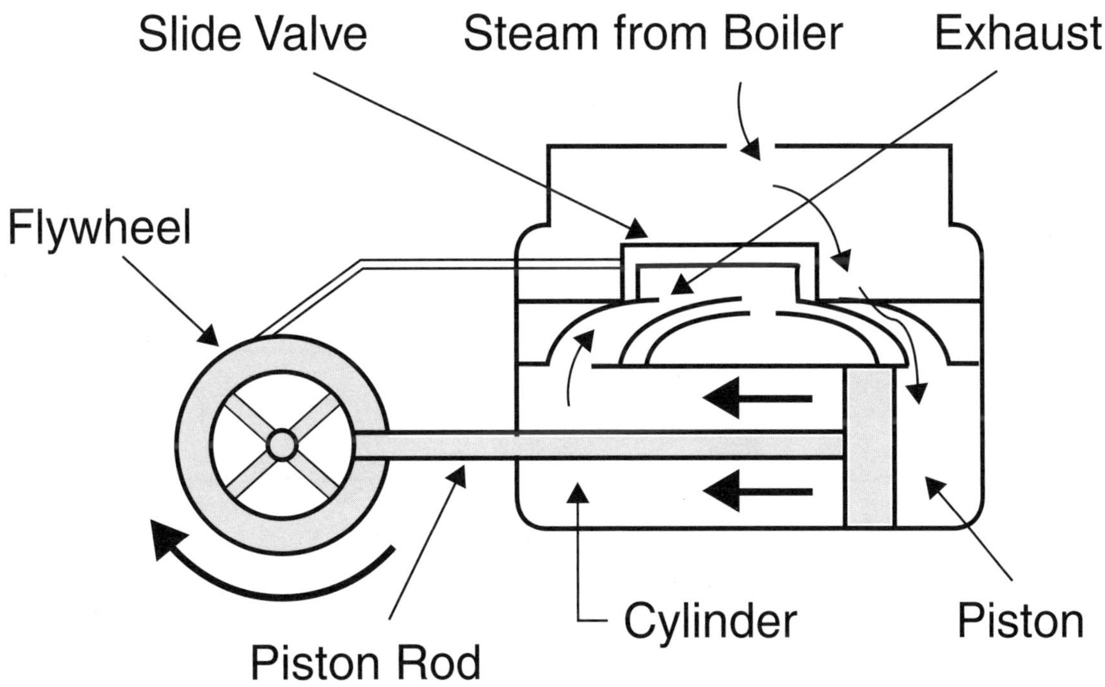

Electric Motor

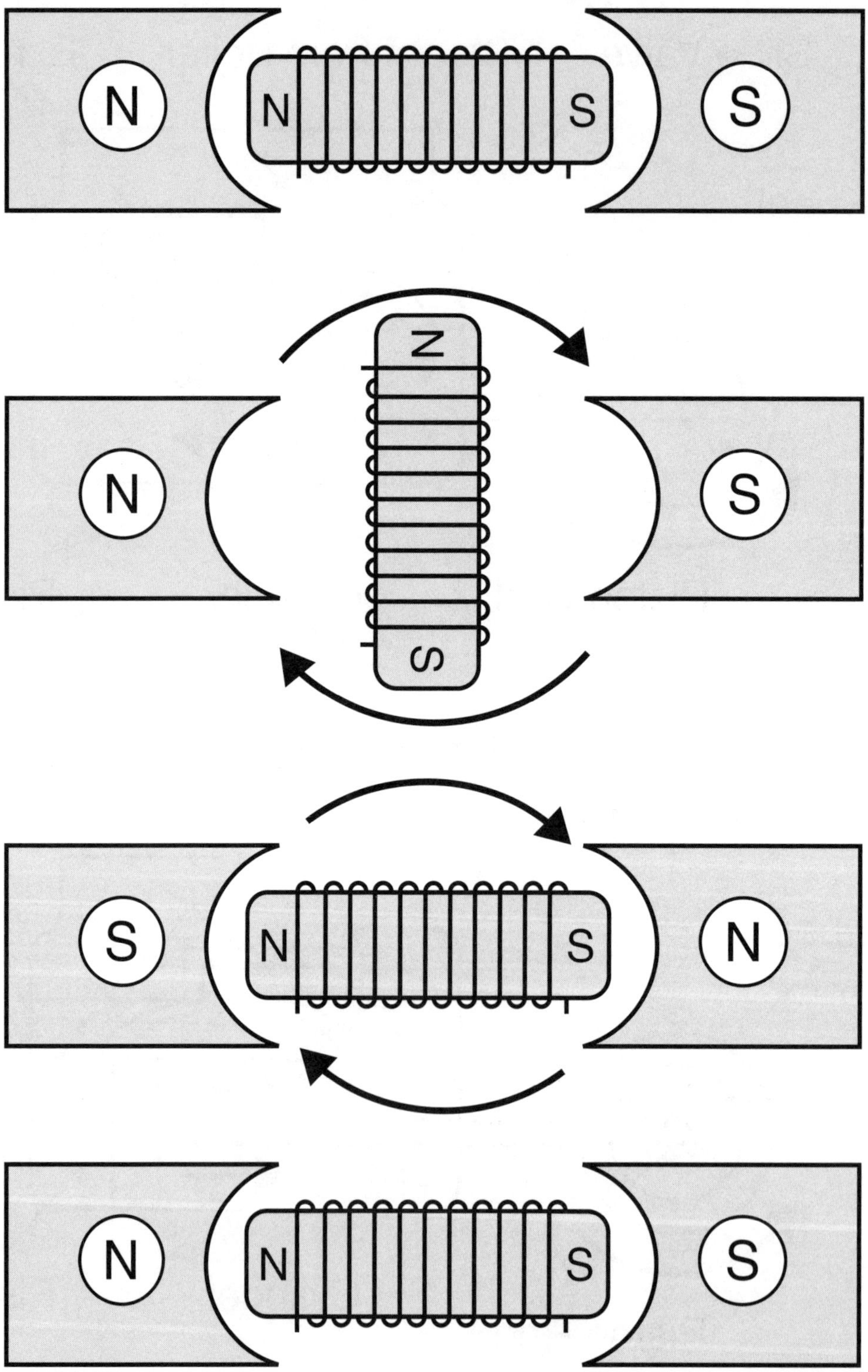

Parts of a Computer

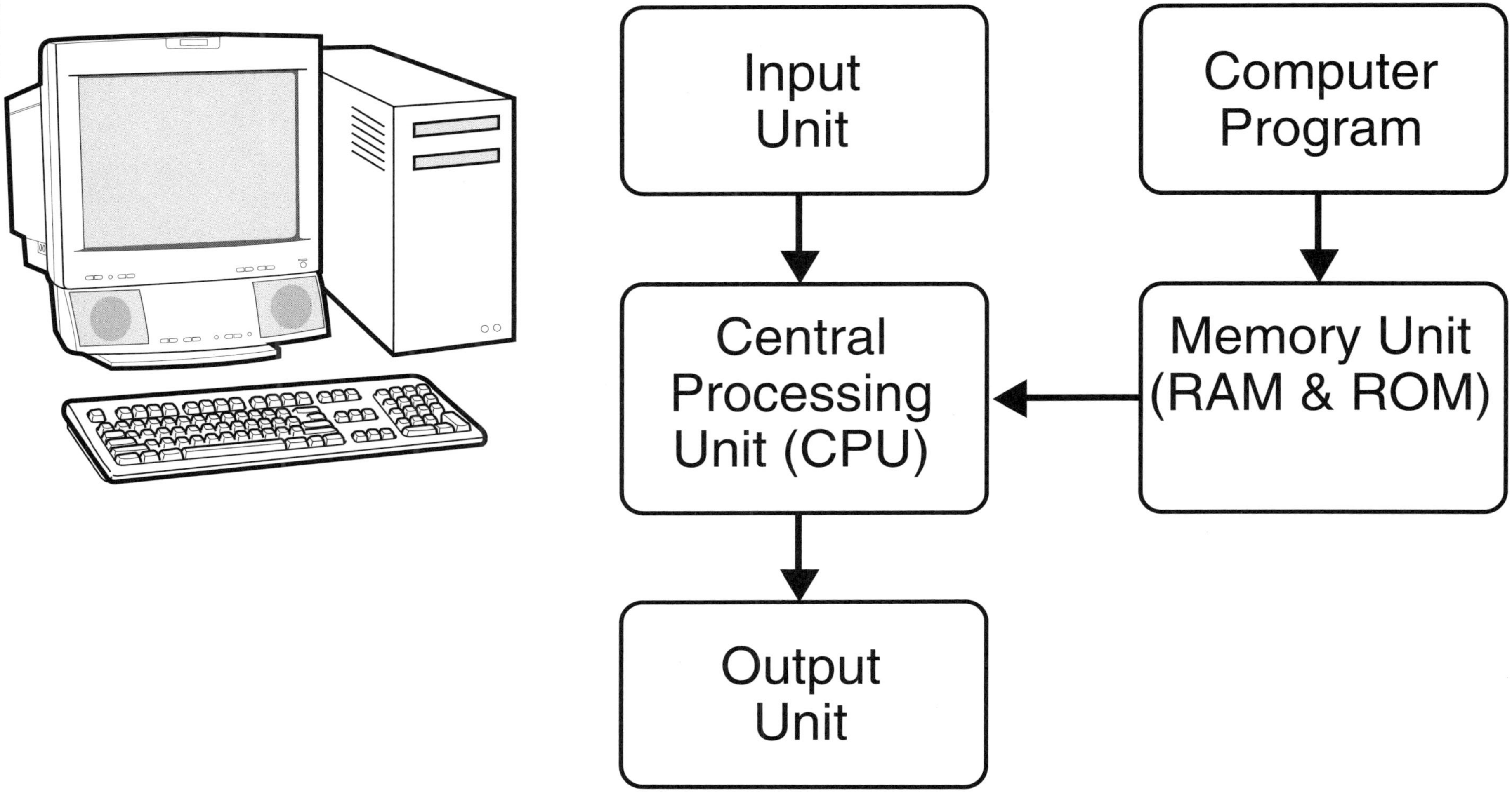

Chapter 8
Measurement Systems and Tools and Their Role in Technology

Objectives

After studying Chapter 8, the students will be able to:

- Define measurement.
- Compare the U.S. customary and the metric measurement systems.
- Describe the major physical qualities that can be measured.
- Name types of measurement.
- Identify common measuring tools to measure linear distances, diameters, and angles.
- Relate measurement to quality control.

Answers to Test Your Knowledge *Questions, Text Pages 172–173*

1. Measurement is the practice of comparing the qualities of an object to a standard.
2. Student answers may include either of the following: There is a logical progression from smaller units to larger ones throughout the system because the base unit always decreases or increases by factors of ten. The metric system is also simpler because it uses a prefix to show us how the base unit is being changed.
3. Volume
4. The common units for length or distance in the metric system are millimeter, centimeter, decimeter, meter, and kilometer. Other units may be accepted at the instructor's discretion.
5. Capacity is the amount of a substance an object can hold.
6. Mass describes the quantity of matter present in an object. Weight is the force of the earth's pull on the mass. Weight therefore can vary depending on gravity.
7. Degree
8. Second
9. Evaluate individually.
10. Evaluate individually.
11. Rule
12. True
13. True
14. Evaluate individually.
15. The process of setting standards, measuring features, comparing them to the standards, and making corrective actions is called quality control. Measurement is one means of determining if a resource fulfills its requirements.

Answers to the Chapter Quiz, Teacher's Resource

1. True
2. False
3. True
4. True
5. True
6. False
7. False
8. False
9. c. Mass
10. d. None of the above
11. a. Linear dimensions
12. c. Protractor
13. b. Weight
14. One is the U.S. customary system, also referred to as the English system or conventional system. The more widely used system is called the International System of Units. This system is abbreviated SI (from the French name *Système International d'Unités*). SI is more commonly known as the metric system.
15. Size and space, mass, time, temperature, number of particles, electrical current, and light intensity

Caption for Reproducible Master

Reproducible Master 8-1. Measurement—Measurement tools can be used to determine linear sizes, diameters, or the angle of adjacent surfaces.

Caption for Transparency Master (Binder/CD Only)

Transparency Master 8-1. Common Factors That Are Measured—Shown are common factors that are used to describe parts and products.

Measurement Systems and Tools and Their Role in Technology

Name ______________________________ Period ______________

Date ______________________________ Score ______________

Chapter 8 Quiz

True/False: If a statement is true, write *True* in the blank. If a statement is false, write *False* in the blank.

_________ 1. Several distance measurements may be used when describing a part or product.

_________ 2. The U.S. customary system of measurement is a decimal system.

_________ 3. Precision measurements are generally not required when building a house.

_________ 4. All measurements compare the feature being measured to a standard.

_________ 5. Machinist's rules are designed for finer measurements than bench rules.

_________ 6. A sliding T-bevel allows for a direct reading when measuring an angle.

_________ 7. Only outputs of a system are measured for quality control.

_________ 8. The measurement tools an operator manipulates and reads are called indirect-reading measurement tools.

Multiple Choice: Choose the best answer and write the corresponding letter in the blank.

_____ 9. Which of the following measures the amount of matter present in a material?
a. Volume
b. Weight
c. Mass
d. Area

_____ 10. On the Fahrenheit scale, what is the number of degrees between the solid point and steam point of water?
a. 100 degrees
b. 212 degrees
c. 32 degrees
d. None of the above

_____ 11. Micrometers are used to measure:
a. linear dimensions.
b. angles.
c. Both a. and b.
d. Neither a. nor b.

_____ 12. Angles can be measured with a:
a. tape rule.
b. micrometer.
c. protractor.
d. All of the above.

(Continued)

Name ____

_____ 13. The gravitational pull exerted on an item is described as:
a. mass.
b. weight.
c. volume.
d. area.

Short Answer: In the space provided, write a brief answer to each of the following questions.

14. List the measurement systems used in the world today.

15. Name the seven physical qualities for which standards have been developed.

Measurement

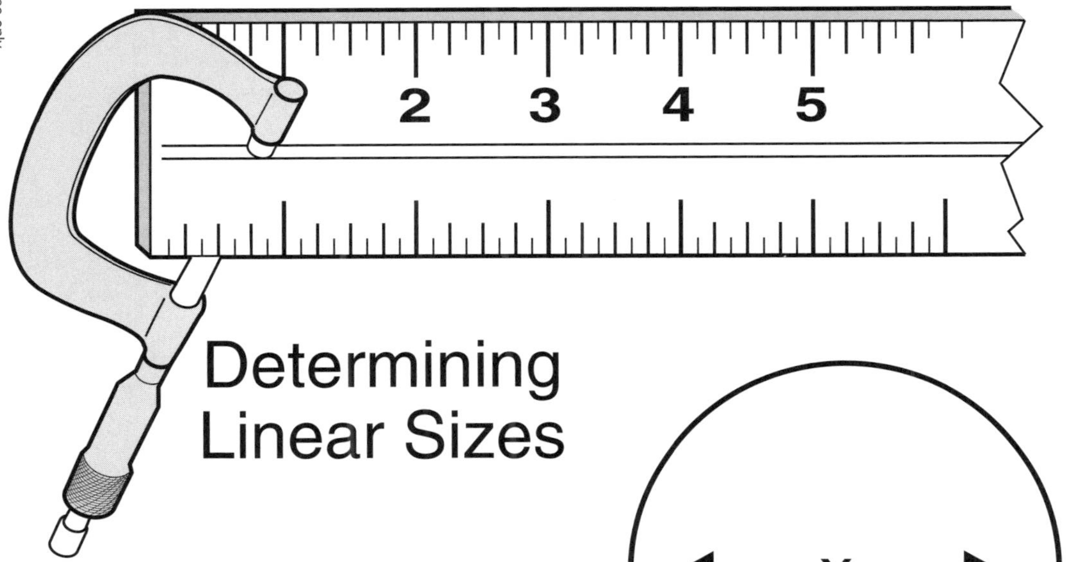

Determining Linear Sizes

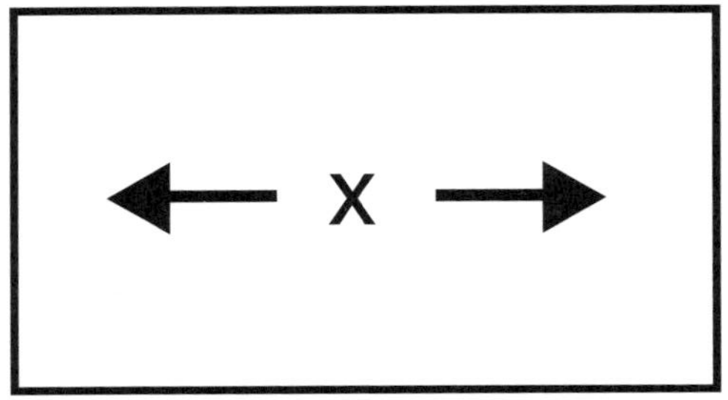

Distances

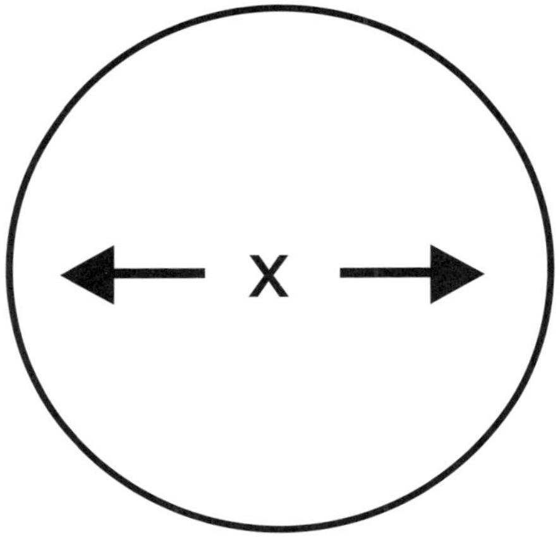

Diameter

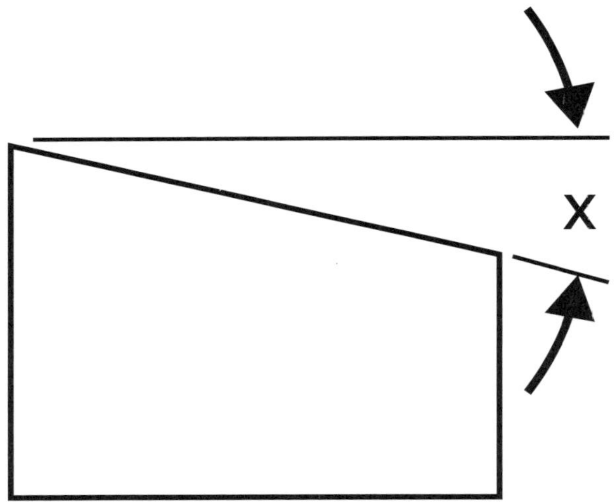

Angles

Section 4

Problem Solving and Design in Technology

Teaching Materials

Text

Chapter 9, pp. 178–189
- Test Your Knowledge, p. 188
- Applying Your Knowledge, p. 189

Chapter 10, pp. 190–204
- Test Your Knowledge, pp. 203–204
- Applying Your Knowledge, p. 204

Chapter 11, pp. 206–222
- Test Your Knowledge, p. 219
- Applying Your Knowledge, p. 220

Chapter 12, pp. 224–241
- Test Your Knowledge, pp. 238–239
- Applying Your Knowledge, p. 239

Section Activities
- Activity 4A—Design Problem, p. 242
- Activity 4B—Fabrication Problem, pp. 242–243

Student Activity Manual

Activity 4A/4B—Design/Fabrication Problem, pp. 33–38

Taking Your Technology Knowledge Home, pp. 39–42

Teacher's Resources

Chapter 9 Quiz, pp. 126–127
Chapter 10 Quiz, pp. 133–134
Chapter 11 Quiz, pp. 140–141
Chapter 12 Quiz, pp. 144–145

Reproducible Masters
- 9-1: Phases of the Technology Development Process, p. 128
- 9-2: Technological Design Criteria, p. 129
- 9-3: Information Needed for Technological Design, p. 130
- 10-1: Types of Design, p. 135
- 10-2: Product Design Process, p. 136
- 10-3: Making an Oblique Drawing, p. 137
- 11-1: Analyzing Designs, p. 142
- 12-1: Types of Engineering Drawings, p. 146

Transparencies (Binder/CD only)
- 9-1: Types of Design Criteria
- 10-1: Types of Product Design
- 11-1: Types of Models
- 12-1: Types of Engineering Drawings

Organization of Section 4

Technology has two major components: a way to develop new knowledge or artifacts and an accumulated body of knowledge about technological practices. We can categorize the latter as knowledge about the following systems: agriculture and related biotechnologies, communication and information, construction, energy and power, manufacturing, medicine, and transportation. The students will explore these types of knowledge in later sections. This section will allow them to investigate the first component, which is often called the problem-solving or design method. It involves four major phases, as shown in Figure 4-1:

- Identifying or recognizing a technological problem
- Developing a solution
- Evaluating the solution
- Communicating the solution

During the study of these major steps, the students first will identify and define a problem. They will then use various sketching techniques to record, evaluate, and modify

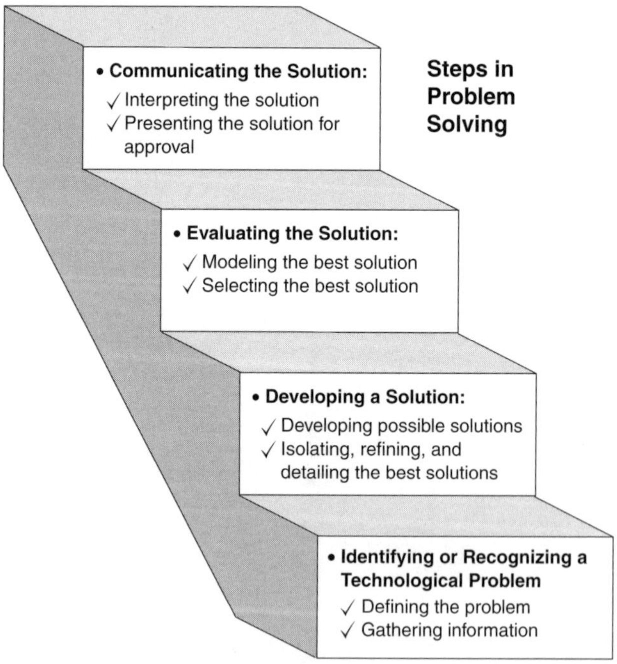

Figure 4-1

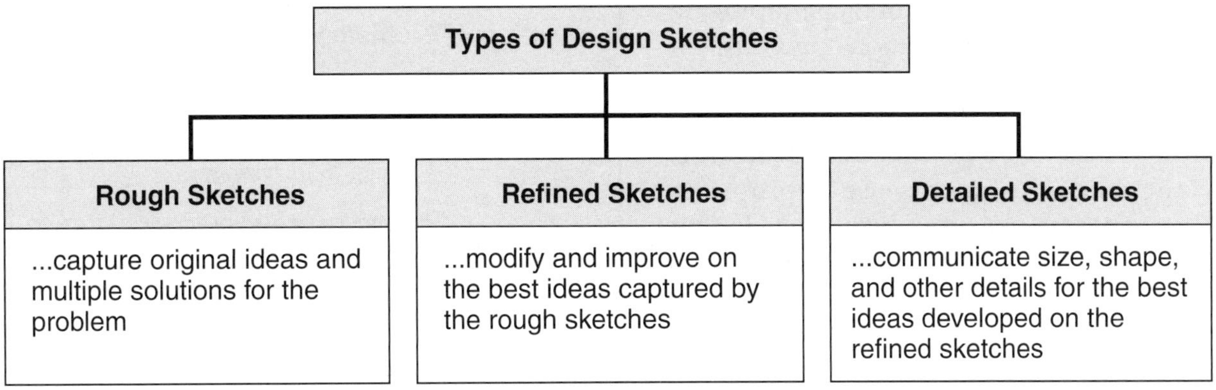

Figure 4-2

product or structure ideas. As shown in Figure 4-2, they will:

- Use rough sketching techniques to generate multiple product ideas
- Develop refined sketches to modify and improve promising ideas
- Generate dimensioned sketches to communicate the best ideas that have been developed

Students will improve and evaluate the ideas using various models. The techniques that the students will be introduced to are:

- Graphic models
- Mathematical models
- Physical models

In addition, the students will use brainstorming techniques to gather suggestions for improving any product ideas they are developing. Section 4 is designed to address the major topics involved in designing technological products. It will include content from four chapters and one or more laboratory activities. The chapters are:

- Chapter 9—The Problem-Solving and Design Process
- Chapter 10—Developing Design Solutions
- Chapter 11—Evaluating Design Solutions
- Chapter 12—Communicating Design Solutions

As with other sections in this program, two different types of activities are provided. Teachers wanting students to address the content from a problem-solving approach should select Activity 4A—Design Problem, in which the students will use a design brief to direct their activities. A more structured laboratory activity in which students develop a bookend is presented by Activity 4B—Fabrication Problem.

Presenting Section 4

Estimated length—5 weeks

Section 4 is designed to help students use the problem-solving method as their new technological devices and systems are developed. They will explore the design processes in four major stages.

First, the students will develop a design statement or problem definition. Second, they will develop several product ideas that meet the problem or opportunity statement. This activity will include rough and refined sketching and brainstorming activities.

The third phase will involve the students developing specifications for the products. This phase will primarily include producing engineering drawings and bills of materials. Finally, the students will present their product for approval.

The content of Section 4 is best presented through short and frequent discussions and demonstrations followed by significant laboratory activities. At each step, the students will be moving a product idea toward a specified, approved product ready for production.

A 25-day schedule, shown in Figure 4-3, has been developed for this section. During this time you will notice that a number of major topics are introduced through presentations and discussions. In addition, the students will use simple sketching and drawing techniques to capture and develop ideas for a new product. **Note:** It is important that the students use appropriate sketching and drawing procedures. Developing advanced drawing skills is beyond the scope of this section, however.

In preparation for presenting this section, you should:

1. Read Chapters 9, 10, 11, and 12 in *Technology.*
2. Prepare discussions on:
   a. Problem solving and design in technology (Chapter 9).
   b. Describing the situation (Chapter 9).
   c. Finding information (Chapter 9).
   d. Developing solutions (Chapter 10).
   e. Brainstorming (Chapter 10).
   f. Modeling solutions (Chapter 11).
   g. Specifying products (Chapter 12).
   h. Bills of materials (Chapter 12).
   i. Presenting products for approval (Chapter 12).
3. Prepare demonstrations on:
   a. Sketching (rough and refined—pictorial).
   b. Dimensioning.
   c. Model building.
   d. Multiview sketching or drawing.
4. Select a laboratory activity.
   a. Activity 4A—This is an open-ended problem-solving activity—designing a travel game.
   b. Activity 4B—This is a structured activity in which groups of students use design techniques to develop a bookend.
5. Prepare teaching materials (transparencies or bulletin boards, for example) to support the content.
6. Determine the study questions and quiz questions that will be used.

Presenting the Activities

Consider the following recommendations in presenting the activity you select for this section.

Section 4: Problem Solving and Design	
Day	**Activity**
1	Discussion: Problem Solving and Design in Technology Introduce laboratory activity
2	Discussion: "Describing the Problem" (Chapter 9) Activity: Problem definition
3	Activity: Developing product criteria
4–5	Discussion: "Finding Information" (Chapter 9) Activity: Researching information
6	Discussion: "Developing Solutions" (Chapter 10) Demonstration: Sketching
7–8	Activity: Rough and refined sketches
9	Discussion: "Brainstorming" (Chapter 10) Activity: Brainstorming
10	Activity: Detailed (dimensioned) sketches Demonstration: Dimensioning
11	Activity: Detailed sketches
12	Discussion: "Modeling Solutions" (Chapter 11) Demonstration: Using machines for producing models
13–17	Activity: Producing a physical model
18	Discussion: "Specifying Products" (Chapter 12) Demonstration: Multiview drawings or sketches
19–21	Activity: Engineering drawing
22	Discussion: "Bills of Materials" (Chapter 12) Activity: Bill of materials
23–24	Discussion: "Presenting Final Product Plans for Approval" (Chapter 12) Activity: Complete products plans
25	Activity: Present product for approval

Figure 4-3

Activity 4A—Design Problem

This activity contains four phases, which correspond with the four chapters covered by the section.

Phase 1: Identifying or recognizing a technological problem (Days 1–5)

Day 1

1. Introduce the design brief to the class: Design a travel game that uses pegs and a 3/4" × 3½" × 3½" wood board. Dice can be used in the game but are not required. Your boss expects you to produce rough sketches for five different ideas, a refined sketch for the best idea, a prototype, and a detail drawing for the board.
2. Divide the class into groups of two to three students each.

Day 2

3. Have each member of the group:
   a. Develop a clear definition for the problem/opportunity.
   b. Present her or his definition to other members of the group.
4. Have each group combine elements of the several definitions developed by members of the group to produce a definition that the group will use.

Day 3

5. Have each group develop engineering, production, marketing, financial, and environmental criteria they believe their product must meet.

Days 4–5

6. Have the members of the group review such items as catalogs and magazines to review a number of solutions that other people developed to answer their design problem.

Phase 2: Developing a solution (Days 6–11)

Days 6–8

1. Have each student:
   a. Develop rough sketches of at least three solutions that address the product definition and meet the product criteria.
   b. Select the most promising solution and prepare a refined sketch of it.
   c. Present the refined sketches to the group.

Day 9

2. Have each group:
   a. Select the most promising design idea from the several produced by its members.
   b. Brainstorm ways to improve the idea.

Days 10–11

3. Have each member produce a dimensioned sketch of the selected and improved product idea.

Phase 3: Evaluating the solution (Days 12–17)

Days 12–17

1. Have each group develop an appearance model or a prototype of its solution.

Phase 4: Communicating the solution (Days 18–25)

Days 18–21

1. Have each group develop a set of multiview engineering drawings for its product. **Note:** Each student can produce a different detail (part) drawing and an assembly drawing for the product.

Day 22

2. Have each member of the group prepare a bill of materials for the product.

Days 23–25

3. Have each group:
   a. Prepare a formal presentation of its product.
   b. Present its product idea to the class.

Activity 4B—Fabrication Problem

This activity is designed for students who need more direction and supervision. The students should follow the same four phases and the major steps outlined in 4A—Design Problem. They need more specific demonstrations, however, and should follow the procedures that are contained in the laboratory activity description in their textbooks.

Reinforcing and Enriching

The following activities can be used to supplement and enrich the laboratory activities at the end of each textbook section and the Applying Your Knowledge activities included at the end of each textbook chapter.

1. Invite a product designer or architect to speak to the class about the process he or she uses to develop a new product or structure.
2. Take the class on a field trip to a company that uses CAD (computer-aided design) to design products and structures.

Taking Your Technology Knowledge Home

This take-home activity allows students to apply their knowledge of problem solving and design to devise a means of sending coded messages to a friend. The code would be transmitted by either visual methods (such as flags or flashes of light) or electronic means (such as telegraph or radio). To complete the activity, the students will:

1. Decide on the method to be used for transmitting the message.
2. Devise a substitution code (letter, number, or symbol) for each letter of the alphabet and the numbers 0–9.
3. Provide a copy of the code to the friend who will receive the message so that the message can be decoded.
4. Sketch the technical system to be used for transmitting the code. Examples of such sketches might be the electrical circuit for a telegraph or flashing light system or the construction details for flags used in a semaphore system.
5. Construct and test the sketched transmission system.
6. Encode the test phrase "Can you read this?"
7. Transmit the encoded phrase to a friend.
8. Receive and translate (decode) the friend's encoded response.

Adjustments for Exceptional Students

The two laboratory activities included in the textbook are designed to accommodate students of all ability levels. The design problem will be challenging for gifted students, who may be able to develop a broader range of creative answers. Lower achievers can still solve the design problem but with solutions within their ability level. The fabrication problem is appropriate for students who need more structure and guidance. Whichever activity is chosen, the amount of structure and teacher guidance should be adjusted to meet students' needs. Students should be given only as much structure as they require for a successful experience, however. The fabrication problem includes considerable drawing and sketching. It should be used as a communication tool and not as a skill in its own right. The course is not designed to produce drafters but to acquaint students with the role of drawing in designing products and structures and how these designs are communicated. Throughout the course, group work allows students to develop cooperative work habits and to contribute their own unique abilities to the team.

Evaluation Procedures

The understanding students have of technology and technological systems should be measured in several ways, including:

- Observation and critical listening during class presentations and discussions.
- Worksheets and quizzes covering the cognitive material presented.
- Open-ended questions about the learned material and how technology is developed.
- Assessment of individual participation and contributions to group work.
- Assessment of the quality and creativeness of the solutions to the assigned design/fabrication activities.

Chapter 9
The Problem-Solving and Design Process

Objectives

After studying Chapter 9, the students will be able to:

- Describe the general problem-solving/design process.
- Define a technological problem or opportunity.
- Describe the technological problem-solving/design process.
- List the major steps involved in solving technological problems and meeting technological opportunities.
- Describe the steps followed in identifying a technological problem or opportunity.
- List the types of criteria and constraints to be considered in the problem-solving/design process.
- Describe the types of information gathered as a foundation for technological development projects.
- List the general methods of gathering information for the problem-solving/design process.

Answers to Test Your Knowledge *Questions, Text Page 188*

1. False
2. Develop an understanding of the problem through observation and investigation. Devise a plan for solving the problem. Implement the plan. Evaluate the plan.
3. You face a technological problem if you need to develop tools, machines, or systems to help you do work. If a "nonproblem" condition can be improved with technology, a technological opportunity exists.
4. Identifying or recognizing a technological problem. Developing a solution. Evaluating the solution. Communicating the solution.
5. Describe the situation that needs a technological solution.
6. True
7. Evaluate individually.
8. Evaluate individually.
9. Historical information, scientific information, technological information, human information, legal information, and ethical information.
10. Historical method, descriptive method, and experimental method.

Answers to the Chapter Quiz, Teacher's Resource

1. False
2. True
3. True
4. False
5. True
6. a. Defining the problem
7. c. Marketing criteria
8. d. Ethical information
9. b. Descriptive methods
10. Identifying or recognizing a technological problem. Developing a solution. Evaluating the solution. Communicating the solution.

Captions for Reproducible Masters

Reproducible Master 9-1. Phases of the Technology Development Process—Shown are the four phases that take problems or opportunities and turn them into solutions.

Reproducible Master 9-2. Technological Design Criteria—Listed are the five design criteria that devices and systems must meet.

Reproducible Master 9-3. Information Needed for Technological Design—Listed are six types of information that are used as a foundation for technological development activities

Caption for Transparency Master (Binder/CD Only)

Transparency Master 9-1. Types of Design Criteria—Shown are the five types of criteria that are used to evaluate designs.

The Problem-Solving and Design Process

Name ______________________________ Period ____________

Date ______________________________ Score ____________

Chapter 9 Quiz

True/False: If a statement is true, write *True* in the blank. If a statement is false, write *False* in the blank.

_________ 1. All technology is designed to solve a problem.

_________ 2. Operational and safety characteristics of a product are based on how, where, and by whom the product will be used.

_________ 3. Technological problem solving replaces a labor-intensive process with an equipment-intensive process.

_________ 4. Technological information includes natural laws and principles that must be considered in developing the solution.

_________ 5. Communicating a technological solution to a problem includes obtaining proper approval.

Multiple Choice: Choose the best answer and write the corresponding letter in the blank.

_____ 6. The first step in solving technological problems involves:
- a. defining the problem.
- b. developing possible solutions.
- c. modeling the best solution.
- d. interpreting the solution.

_____ 7. Which of the following describes the functional characteristics a customer wants in a device?
- a. Technical criteria
- b. Financial criteria
- c. Marketing criteria
- d. Production criteria

_____ 8. Which of the following types of information describes the values people have toward similar devices and systems?
- a. Human information
- b. Legal information
- c. Technological information
- d. Ethical information

_____ 9. What type of information-gathering process records observations of present conditions?
- a. Historical methods
- b. Descriptive methods
- c. Experimental methods
- d. Both b. and c.

(Continued)

Name ______________________________

Short Answer: In the space provided, write a brief answer to the following question.

10. List the four major steps in the problem-solving method used to develop new technology.

____

____

____

____

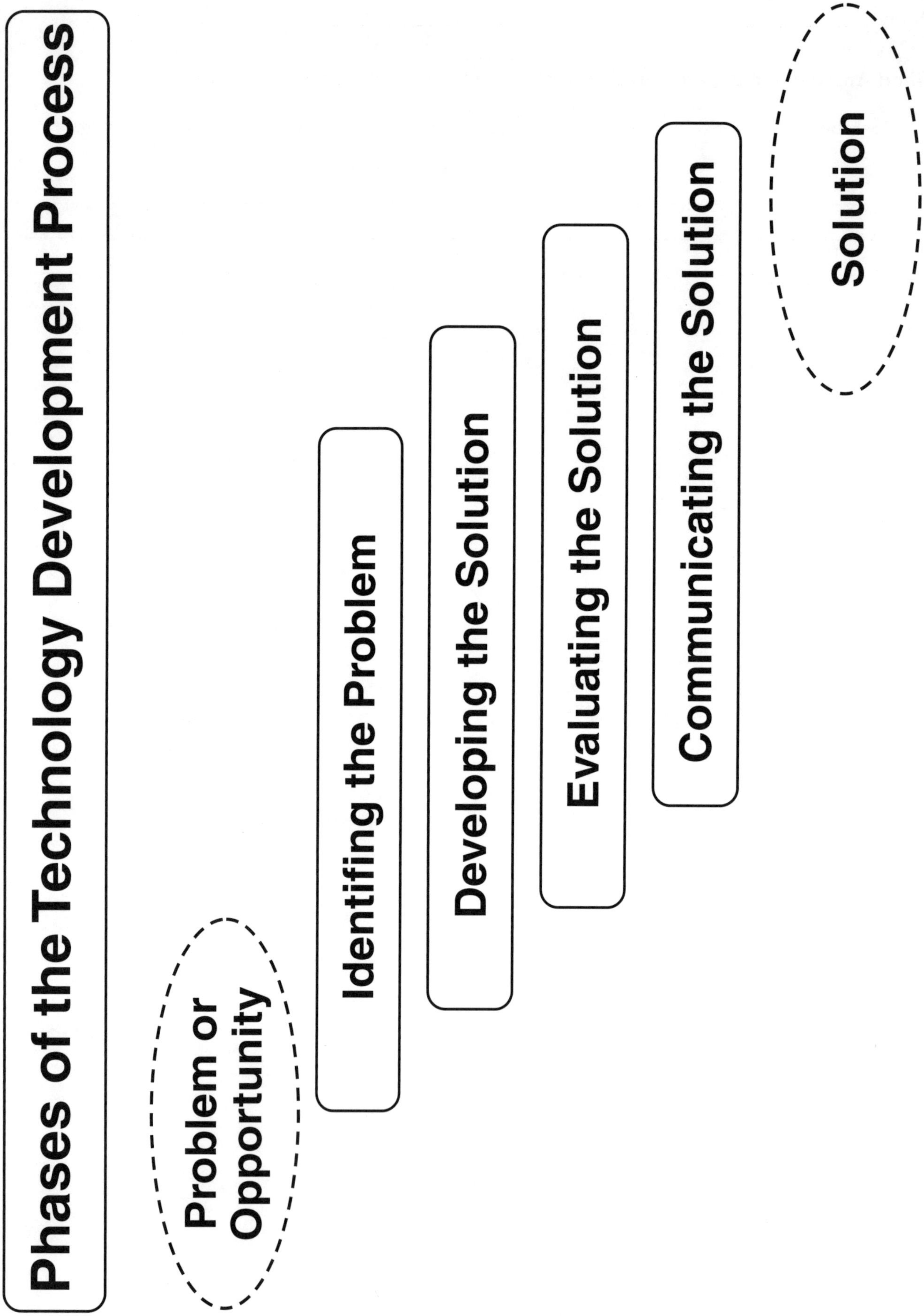
Phases of the Technology Development Process
Problem or Opportunity
Identifing the Problem
Developing the Solution
Evaluating the Solution
Communicating the Solution
Solution

Technological Design Criteria

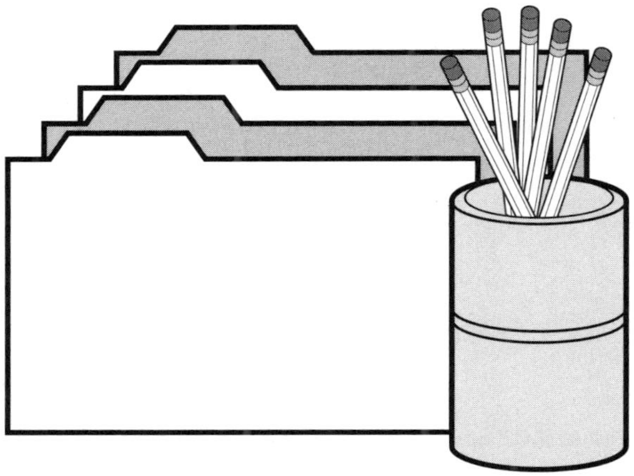

- ✓ Technical Criteria
- ✓ Production Criteria
- ✓ Market Criteria
- ✓ Financial Criteria
- ✓ Environmental Criteria

Information Needed for Technological Design

- Technological
- Scientific
- Historical
- Human
- Ethical
- Legal

Chapter 10
Developing Design Solutions

Objectives

After studying Chapter 10, the students will be able to:

- Name the two major areas in which designers develop solutions to technological problems and opportunities.
- List the three main steps followed in developing technological designs.
- Identify the ways in which ideas can be stimulated when developing design solutions.
- Differentiate between kinds of design sketches.
- List the three major types of information required for detailed sketches when building models.
- Identify the three types of pictorial sketches used in product design.

Answers to Test Your Knowledge *Questions, Text Pages 203–204*

1. System design problems and opportunities. Product design problems and opportunities.
2. Developing preliminary solutions. Isolating and refining the best solution. Detailing the best solution.
3. Divergent thinking is used to think of as many solutions as possible. Convergent thinking is used to narrow and focus the ideas until the most feasible solution is found.
4. False
5. Student answers may include any of the following: Encourage wild, far-out ideas. Record the ideas without reacting to them. Seek quantity, not quality. Keep up a rapid pace.
6. Classification involves dividing the problem into increasingly specific categories.
7. A "what if" scenario is a wild hypothetical situation that can be used to help develop solutions.
8. Rough sketches are sketches of designs that are incomplete and unrefined. Refined sketches are sketches of designs that have been improved through study, contemplation, or the merging of ideas. Detailed drawings communicate the information needed to build a model of the product or structure.
9. Size information, location information, and geometry information
10. b. Isometric
11. c. Perspective
12. a. Oblique
13. b. Isometric
14. a. Oblique
15. c. Perspective
16. c. Perspective
17. b. Isometric
18. a. Oblique
19. The number of vanishing points used. A one-point perspective shows an object as if you were directly in front of it. A two-point perspective shows how an object would appear if you stood at one corner. A three-point perspective shows how the eye would see the length, width, and height of an object.
20. True

Answers to the Chapter Quiz, Teacher's Resource

1. False
2. True
3. True
4. True
5. True
6. False
7. True
8. c. Classification
9. b. The number of ideas generated by individuals working in a group is usually greater than the number generated by the individuals working alone.
10. a. Rough sketches
11. d. Detailed sketch
12. a. Oblique sketches
13. b. Size information
14. System design deals with the arrangement of components to produce a desired result.

Product design deals with developing a product or structure that meets the customer's needs.

15. Function, production, and marketing

Captions for Reproducible Masters

Reproducible Master 10-1. Types of Design—Shown are two separate types of design work: product design and systems design.

Reproducible Master 10-2. Product Design Process—Shown is the circular motion of a product design process with feedback.

Reproducible Master 10-3. Making an Oblique Drawing—Shown are the steps in producing an oblique drawing. Oblique drawings show the front view of an object as if a person was looking directly at it.

Caption for Transparency Master (Binder/CD Only)

Transparency Master 10-1. Types of Product Design—Manufactured products and constructed structures are shown as the two types of product design.

Developing Design Solutions

Name ____ Period ____________________

Date ____ Score ____________________

Chapter 10 Quiz

True/False: If a statement is true, write *True* in the blank. If a statement is false, write *False* in the blank.

_________ 1. When developing preliminary solutions to a technological problem, one must use convergent thinking.

_________ 2. There is not always one "best" solution because consumers have different needs and different amounts of money to spend.

_________ 3. The final design solution may not always be one of the preliminary design solutions.

_________ 4. A two-point perspective drawing is constructed much like an isometric drawing.

_________ 5. System design deals with the arrangement of components to produce a desired result.

_________ 6. When brainstorming, the group should carefully monitor the quality of ideas produced.

_________ 7. Isometric sketches show the product's view from one corner.

Multiple Choice: Choose the best answer and write the corresponding letter in the blank.

_____ 8. Which of the following methods of developing preliminary solutions divides the problem into major segments and subparts?
- a. "What if" scenarios
- b. Brainstorming
- c. Classification
- d. None of the above

_____ 9. As related to brainstorming, the concept of synergism is that:
- a. the sum of ideas generated by people working alone is greater than the sum of ideas generated by individuals working in a group.
- b. the number of ideas generated by individuals working in a group is usually greater than the number generated by the individuals working alone.
- c. individuals will produce the same number of ideas whether they work in a group or alone.
- d. None of the above.

_____ 10. Which type of sketch communicates shape and proportion but not size?
- a. Rough sketch
- b. Detail sketch
- c. Dimensioned sketch
- d. Isometric sketch

_____ 11. Which type of sketch provides the information needed to build a model of the project?
- a. Refined sketch
- b. Rough sketch
- c. Perspective sketch
- d. Detailed sketch

(Continued)

Name ____

_____ 12. Which type of sketch directly shows the front view of a product?
a. Oblique sketch
b. Perspective sketch
c. Isometric sketch
d. Both a. and c.

_____ 13. Which type of information contained in a drawing communicates the dimensions of the product?
a. Geometry information
b. Size information
c. Location information
d. Both a. and b.

Short Answer: In the space provided, write a brief answer to each of the following questions.

14. What is the difference between system design and product design?

____

____

____

____

____

15. List the three factors that a product must be designed to satisfy.

____

____

____

Types of Design

Product Design

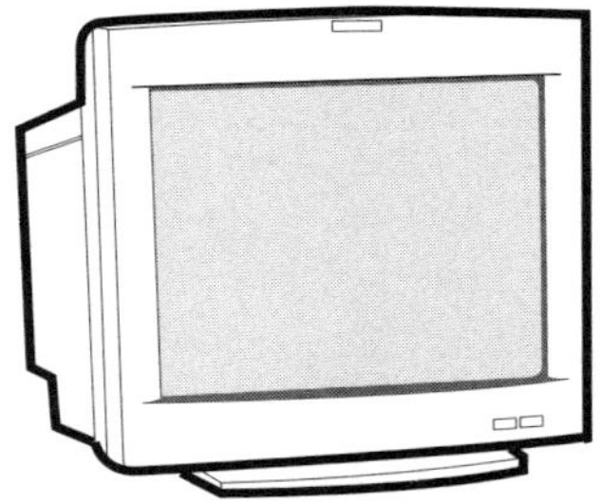

System Design

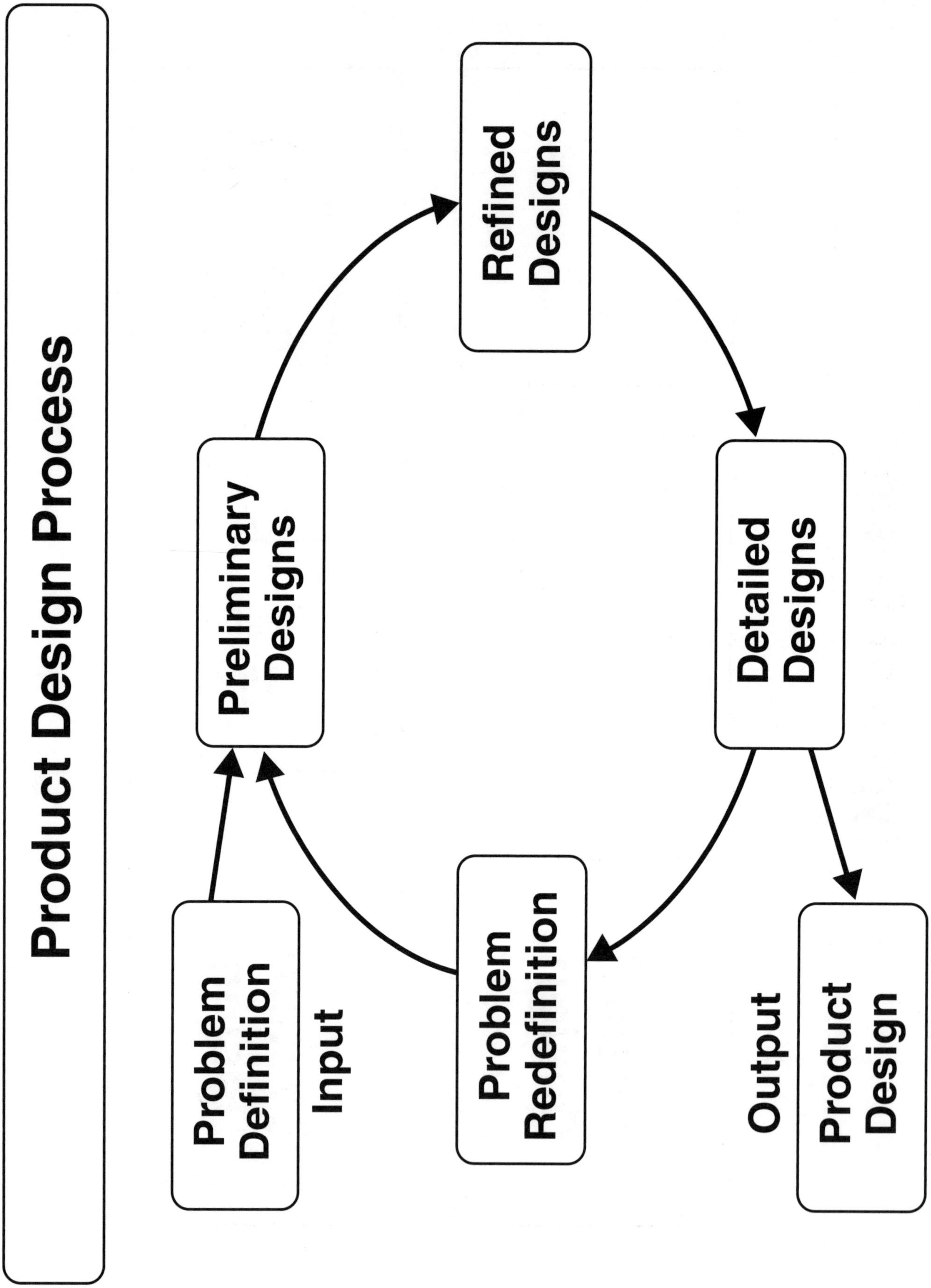
Product Design Process
Problem Definition
Input
Preliminary Designs
Refined Designs
Detailed Designs
Problem Redefinition
Output
Product Design

Making an Oblique Drawing

Establish Front View

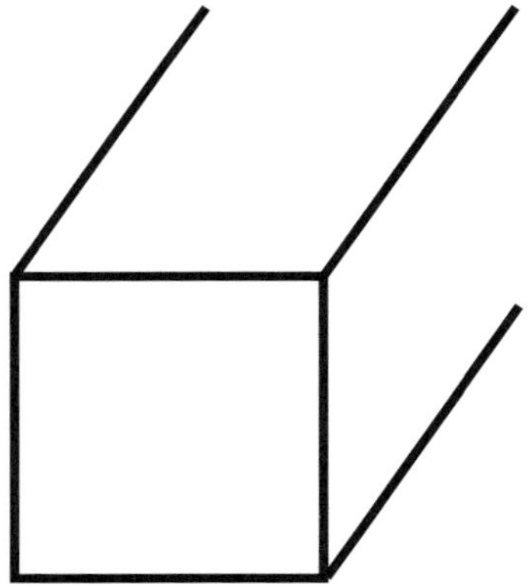

Establish Side and Top View

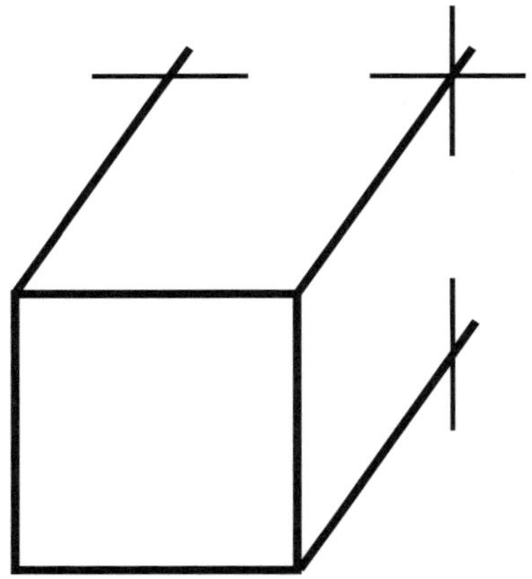

Establish Depth

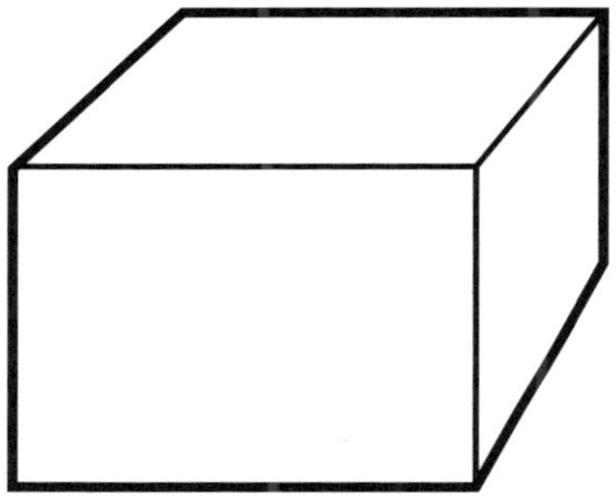

Complete Block Outline

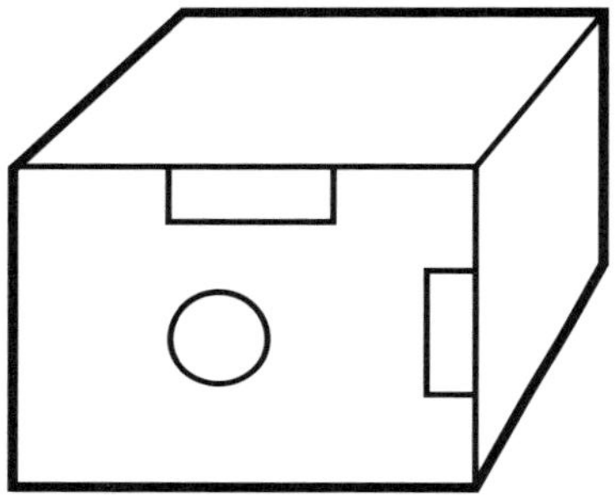

Locate Features

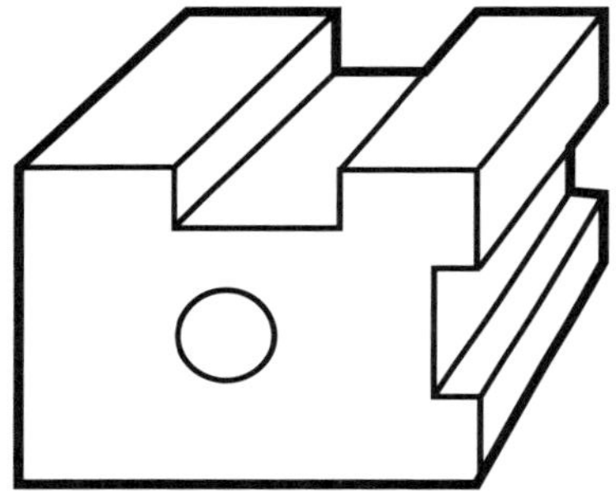

Complete Drawings

Chapter 11
Evaluating Design Solutions

Objectives

After studying Chapter 11, the students will be able to:

- Define modeling.
- Describe how models are used in the design of technological artifacts and systems.
- List the three major types of models.
- Differentiate between types of models.
- Describe the types of analysis used to evaluate designs.
- Identify the role of redesigning.

Answers to Test Your Knowledge *Questions, Text Page 219*

1. Modeling is the activity of simulating actual events, structures, or conditions.
2. Evaluate individually.
3. Graphic models, mathematical models, and physical models
4. Evaluate individually.
5. Prototype
6. Wire frame model, surface model, and solid model
7. Functional
8. Human factors analysis (ergonomics)
9. False
10. True

Answers to the Chapter Quiz, Teacher's Resource

1. False
2. False
3. True
4. False
5. True
6. c. Diagrams
7. b. Mathematical models
8. a. Surface models
9. c. Specification analysis
10. a. Market analysis
11. b. Human factors analysis
12. b. Prototype
13. d. Mock-up
14. Evaluate individually. Answers may include emergence of new technology, changes in the attitudes of people or the availability of resources, and flaws in previous design.
15. Functional analysis, specification analysis, human factors analysis, market analysis, and economic analysis

Caption for Reproducible Master

Reproducible Master 11-1. Analyzing Designs—Listed are the five factors that must be analyzed when designing a product or system.

Caption for Transparency Master (Binder/CD Only)

Transparency Master 11-1. Types of Models—Graphic, physical, and mathematical models are shown as the three types of models used in product design. Computer models are either graphic or mathematical.

Evaluating Design Solutions

Name ______________________________ Period ______________

Date ______________________________ Score ______________

Chapter 11 Quiz

True/False: If a statement is true, write *True* in the blank. If a statement is false, write *False* in the blank.

_________ 1. Graphs show the relationship between components in a system.

_________ 2. A mock-up is a working model of a product.

_________ 3. Economic analysis is used to evaluate the chances for making a profit on a certain product.

_________ 4. High-precision specifications are always required for a product to function properly.

_________ 5. Computer simulation can be used to test the strength of a product.

Multiple Choice: Choose the best answer and write the corresponding letter in the blank.

_____ 6. Which of the following show the relationship between components of a system?
a. Graphs
b. Charts
c. Diagrams
d. Both b. and c.

_____ 7. Which type of models show the relationship of things through formulas?
a. Physical models
b. Mathematical models
c. Graphic models
d. None of the above

_____ 8. Which of the following computer models is frequently used in developing sheet metal products?
a. Surface models
b. Wire frame models
c. Solid models
d. None of the above

_____ 9. Evaluating the materials and tolerances specified for a design is known as:
a. functional analysis.
b. market analysis.
c. specification analysis.
d. human factors analysis.

_____ 10. Studying present and anticipated competition is one aspect of:
a. market analysis.
b. economic analysis.
c. human factors analysis.
d. functional analysis.

(Continued)

Name ____

_____ 11. Ergonomics is a concept that relates to:
- a. market analysis
- b. human factors analysis.
- c. functional analysis.
- d. economic analysis.

_____ 12. Which of the following is a physical model?
- a. Surface model
- b. Prototype
- c. Solid model
- d. Graph

_____ 13. Which of the following is *not* a type of graphic model?
- a. Chart
- b. Conceptual model
- c. Diagram
- d. Mock-up

Short Answer: In the space provided, write a brief answer to each of the following questions.

14. Name two situations that may cause a product to be redesigned.

____

____

15. List the five types of design analysis.

____

____

____

____

____

Analyzing Designs

Designs

- ✓ Function
- ✓ Specifications
- ✓ Human Factors
- ✓ Market Acceptance
- ✓ Economics

Chapter 12
Communicating Design Solutions

Objectives

After studying Chapter 12, the students will be able to:

- List the three basic documents through which designers communicate product information.
- Identify the three types of engineering drawings.
- Describe a computer-aided design system.
- Identify the information found on a bill of materials.
- Identify the information found on a specification sheet.
- Name the types of reports used to gain approval for designed products.

Answers to Test Your Knowledge *Questions, Text Pages 238–239*

1. Engineering documents, bill of materials, and specification sheets
2. c. Systems
3. a. Detail
4. b. Assembly
5. b. Assembly (**Note**: *c. Systems* may also be accepted at the instructor's discretion.)
6. c. Systems
7. c. Systems
8. a. Detail
9. a. Detail
10. b. Assembly
11. Student answers may include any of the following: CAD systems do the job more quickly and uniformly. Also, computer drawings are easier to correct, store, and communicate. Computer drafters can send their drawings across the country or around the world in seconds using the Internet.
12. False
13. True
14. Technical data sheets are lists of specifications prepared by manufacturers for the products they have on the market.
15. Student answers may include: need statements, proposed design solutions, cost estimates, marketing strategies, economic forecasts, and environmental impact statements.

Answers to the Chapter Quiz, Teacher's Resource

1. True
2. False
3. True
4. True
5. True
6. True
7. False
8. False
9. True
10. b. Two-view drawings
11. c. Object lines
12. d. All of the above
13. a. System drawings
14. Engineering drawings, bills of materials, and specification sheets
15. Detail drawings commonly contain all the information needed to manufacture one part. Assembly drawings show how parts fit together to make assemblies. System drawings show how parts in a system relate to each other and work together.

Caption for Reproducible Master

Reproducible Master 12-1. Types of Engineering Drawings—Listed are the three types of engineering drawings. These drawings can be generated by computer or by hand.

Caption for Transparency Master (Binder/CD Only)

Transparency Master 12-1. Types of Engineering Drawings—Shown are the three types of engineering drawings: detail, assembly, and systems.

Communicating Design Solutions

Name ______________________________ Period ______________

Date ______________________________ Score ______________

Chapter 12 Quiz

True/False: If a statement is true, write *True* in the blank. If a statement is false, write *False* in the blank.

_________ 1. Detail drawings usually contain all the information to manufacture all the parts needed for a complete product.

_________ 2. Detail drawing should show as many views as possible.

_________ 3. Tolerance is the acceptable amount of deviation from a given dimension.

_________ 4. Center lines locate holes in a part and should be constructed as a light line with alternating long and short dashes.

_________ 5. Orthographic assembly drawings use a single view to show how parts fit together.

_________ 6. The major benefits of computer-aided design are that it is faster, the drawings are more uniform, and the drawings are easier to correct and store.

_________ 7. The physical properties of a material include strength and hardness.

_________ 8. Technical data sheets are prepared by large purchasers to detail specifications for materials and products they need.

_________ 9. Extension lines indicate the points from which measurements are taken.

Multiple Choice: Choose the best answer and write the corresponding letter in the blank.

_____ 10. Which of the following types of orthographic projections are used to show the size and shape of cylindrical parts?
- a. One-view drawings
- b. Two-view drawings
- c. Three-view drawings
- d. None of the above

_____ 11. Which of the following lines show the shape of the object and should be the darkest lines of the drawing?
- a. Hidden lines
- b. Center lines
- c. Object lines
- d. Both b. and c.

_____ 12. Which of the following information is included on a bill of materials?
- a. The part number
- b. The size of the part
- c. A description of the part
- d. All of the above

(Continued)

Name ____

_____ 13. Which of the following would show the relationship between electrical and hydraulic or pneumatic components?
a. System drawings
b. Detail drawings
c. Assembly drawings
d. None of the above

Short Answer: In the space provided, write a brief answer to each of the following questions.

14. List the three types of documents used to communicate a product's specifications.

____

____

____

15. Describe each of the following: detailed drawings, assembly drawings, and system drawings.

____

____

____

____

____

Types of Engineering Drawings

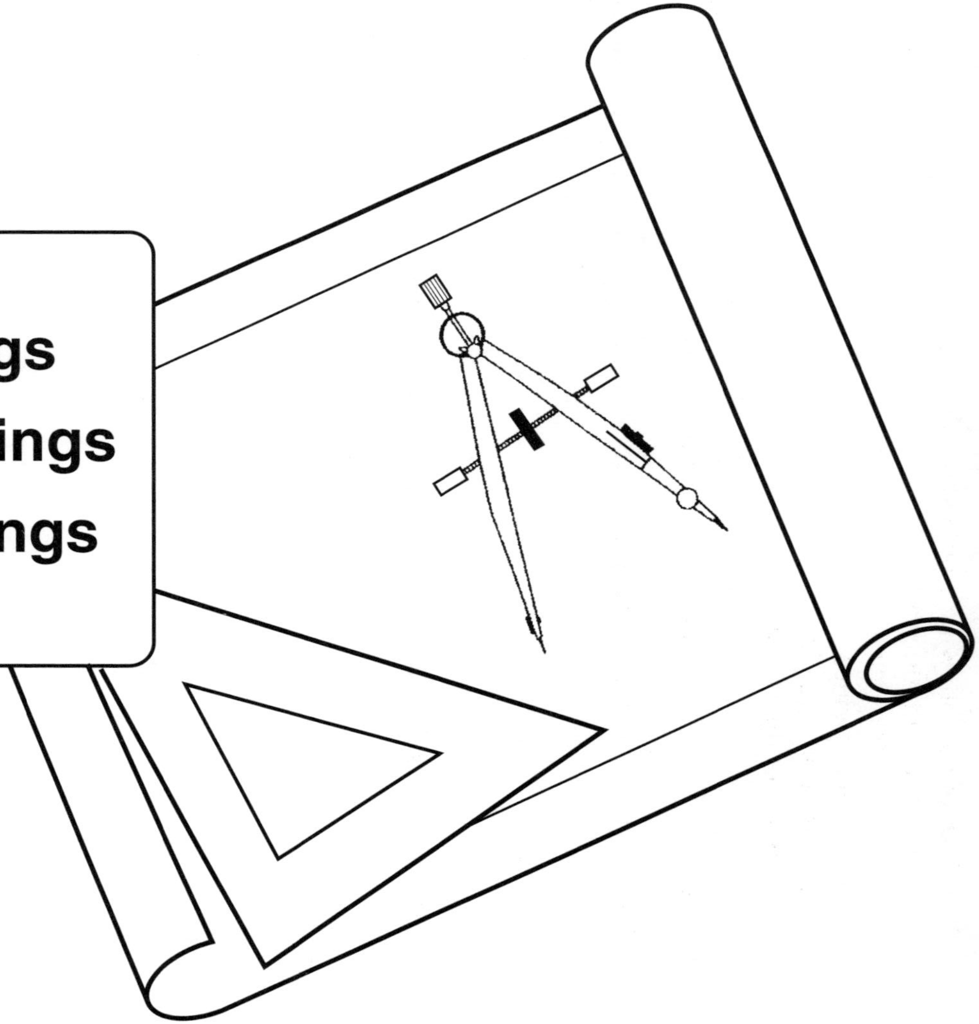

Detail Drawings
Assembly Drawings
Systems Drawings

Section 5

Applying Technology: Producing Products and Structures

13 Using Technology to Produce Artifacts

14 The Types of Material Resources and How They Are Obtained

15 Processing Resources

16 Manufacturing Products

17 Constructing Structures

18 Using and Servicing Products and Structures

Teaching Materials

Text

Chapter 13, pp. 246–253
  Test Your Knowledge, p. 253
  Applying Your Knowledge, p. 253

Chapter 14, pp. 254–269
  Test Your Knowledge, pp. 268–269
  Applying Your Knowledge, p. 269

Chapter 15, pp. 270–288
  Test Your Knowledge, pp. 287–288
  Applying Your Knowledge, p. 288

Chapter 16, pp. 290–317
  Test Your Knowledge, p. 316
  Applying Your Knowledge, p. 317

Chapter 17, pp. 318–343
  Test Your Knowledge, p. 340
  Applying Your Knowledge, p. 341

Chapter 18, pp. 344–355
  Test Your Knowledge, p. 355
  Applying Your Knowledge, p. 355

Section Activities

Activity 5A—Design Problem: Manufacturing Technology, p. 356
Activity 5B—Fabrication Problem: Manufacturing Technology, pp. 356–357
Activity 5C—Design Problem: Construction Technology, p. 358
Activity 5D—Fabrication Problem: Construction Technology, pp. 358–359

Student Activity Manual

Activity 5A/5B—Manufacturing Design/Fabrication Problem, pp. 45–51
Activity 5C—Construction Design Problem, pp. 53–54
Activity 5D—Construction Design Problem, pp. 55–56
Activity 5E—Product Servicing Activity, pp. 57–59
Activity 5F—Papermaking/Paper Recycling Activity, pp. 61–62
Taking Your Technology Knowledge Home, pp. 63–65

Teacher's Resources

Chapter 13 Quiz, pp. 158–159
Chapter 14 Quiz, pp. 163–164
Chapter 15 Quiz, pp. 168–169
Chapter 16 Quiz, pp. 173–174
Chapter 17 Quiz, pp. 183–184
Chapter 18 Quiz, pp. 190–191
Reproducible Masters
  14-1: Manufacturing, p. 165
  15-1: Types of Primary Processes, p. 170
  16-1: Casting and Molding, p. 175
  16-2: Forming, p. 176
  16-3: Separating, p. 177
  16-4: Conditioning, p. 178
  16-5: Assembly, p. 179
  16-6: Finishing, p. 180
  17-1: Construction, p. 185
  17-2: Steps in Constructing a Structure, p. 186
  17-3: Types of Civil Structures, p. 187
  18-1: Steps in Using a Technological Product, p. 192

Transparencies (Binder/CD only)
  13-1: Production Activities
  14-1: Securing Natural Resources
  15-1: Primary Processing
  16-1: Types of Manufacturing
  17-1: Types of Construction
  17-2: Types of Buildings
  18-1: Servicing

Organization of Section 5

Section 5 is structured to allow students to investigate the major actions that are used to obtain and convert natural resources into manufactured products and constructed structures. This study involves four major topics:

- Locating, extracting, and converting natural resources into industrial materials
- Manufacturing products from industrial materials
- Constructing structures from industrial materials and manufactured products
- Using and maintaining products and structures

The section is organized so that the students will explore the various actions that are used to locate and change raw materials into products and structures. This process is shown in Figure 5-1. Next, they will investigate the types of raw materials and the techniques used to locate and

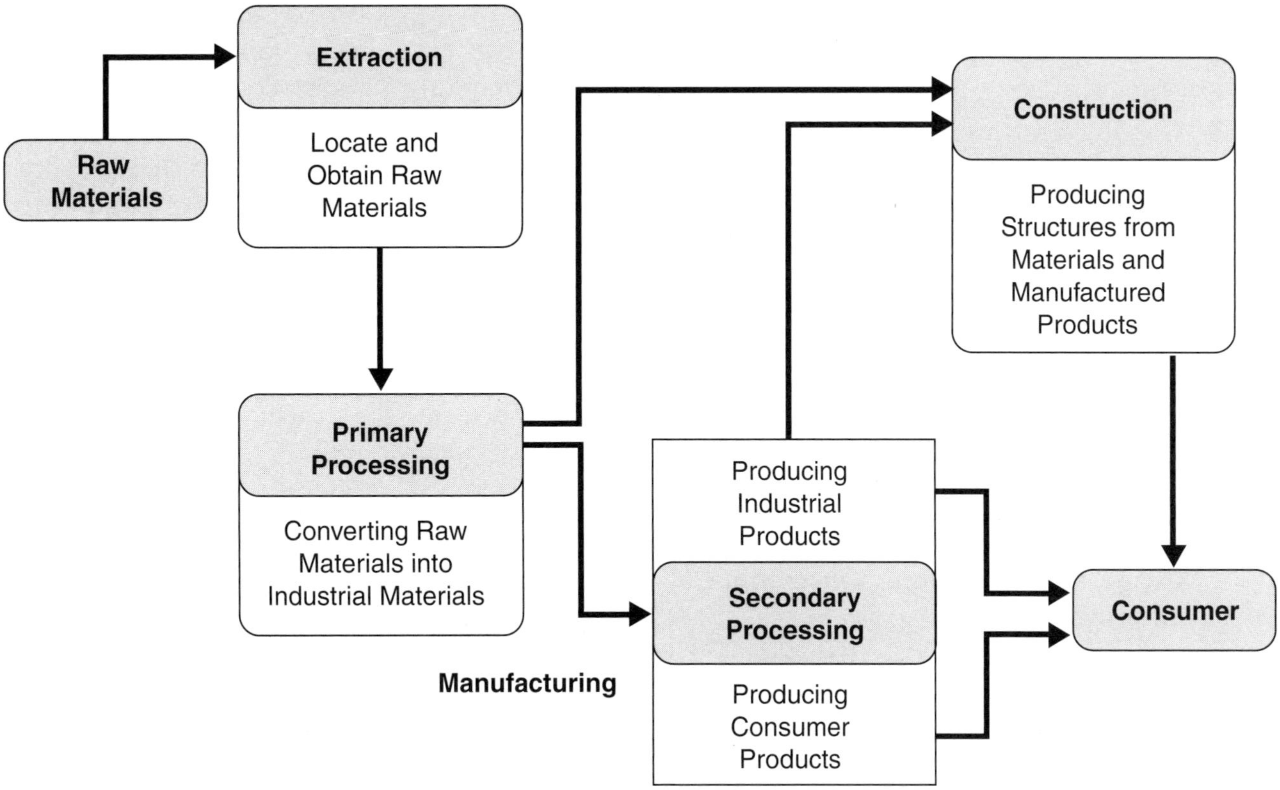

Figure 5-1

obtain them, Figure 5-2. Specifically, they will learn about:

- Genetic materials: Organic materials obtained during the normal life cycle of plants or animals
- Fossil fuel materials: Organic materials made up predominantly of carbon and hydrogen, such as petroleum, natural gas, and coal
- Minerals: Naturally occurring substances with a specific chemical composition, such as ores, nonmetallic minerals, clay, and gems

In addition, the students will explore ways these materials are located and obtained, including harvesting, drilling, and mining.

Next, the students will learn how materials are converted into industrial materials, called standard stock. They will learn about:

- Mechanical processes
- Thermal processes
- Chemical and electrochemical processes

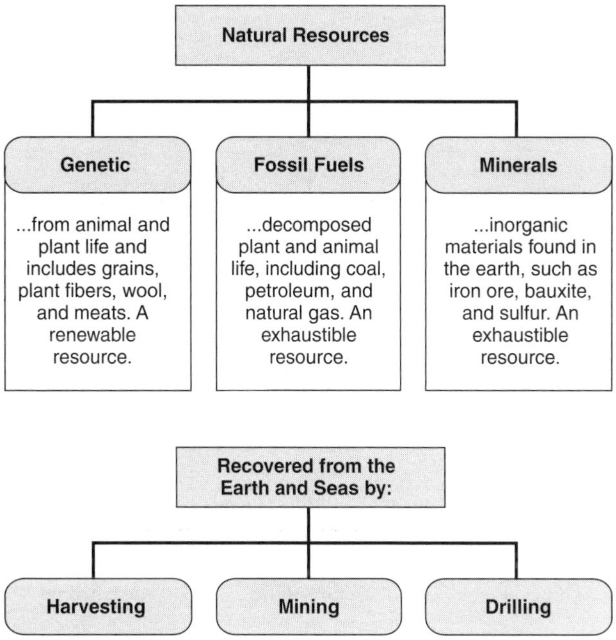

Figure 5-2

Following this study, the students will explore secondary manufacturing, where industrial materials are changed into industrial and consumer products. This type of manufacturing, as shown in Figure 5-3, involves six major types of processes:

- Casting and molding: Processes that change the form of materials by introducing a liquid material into a mold where it is allowed to solidify before being removed
- Forming: Processes that change the form of material by applying a force above the material's yield point and below its fracture point, causing the material to change shape
- Separating: Processes that change a product's size and shape by removing excess material through machining or shearing
- Conditioning: Processes that change the internal structure of a material, and therefore its properties, through thermal, mechanical, and chemical means
- Assembling: Processes that temporarily or permanently hold parts together using mechanical or bonding techniques
- Finishing: Processes that protect or improve the appearance of a product by changing its surface or by applying a coating

Next, the students will study the practices used to construct a structure. The typical steps, as shown in Figure 5-4, are:

- Preparing the site
- Setting foundations
- Building the framework
- Enclosing the structure
- Installing utilities
- Finishing the interior and exterior
- Completing the site

Finally, the students will learn how to select, operate, service, and dispose of technological products. The students will investigate these technological topics through six chapters:

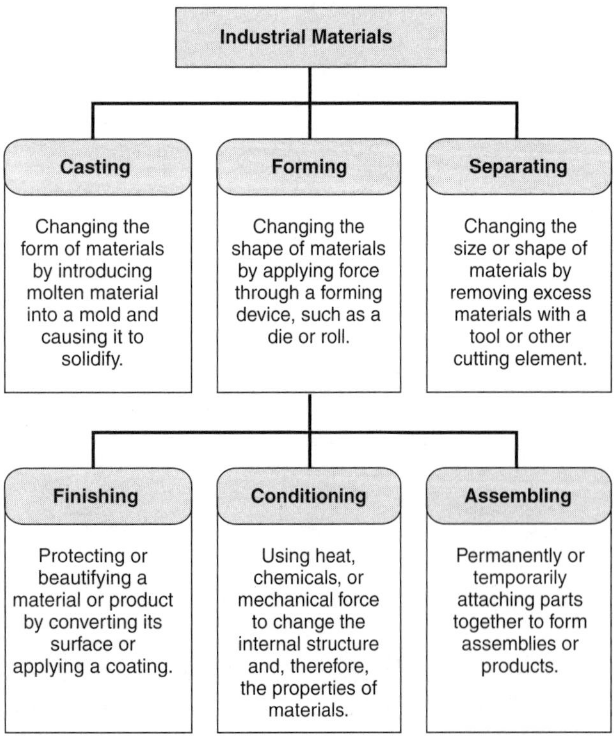

Figure 5-3

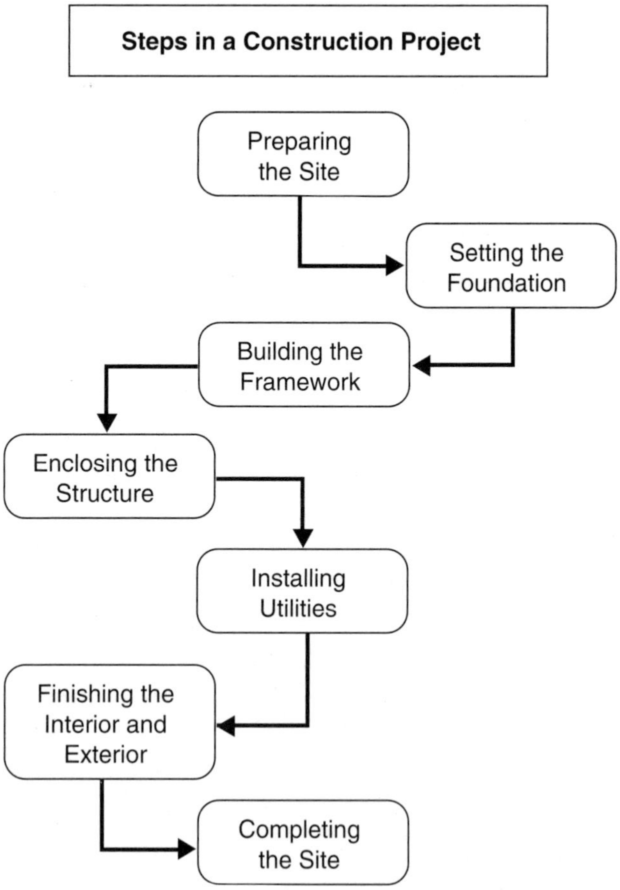

Figure 5-4

- Chapter 13—Using Technology to Produce Artifacts
- Chapter 14—The Types of Material Resources and How They Are Obtained
- Chapter 15—Processing Resources
- Chapter 16—Manufacturing Products
- Chapter 17—Constructing Structures
- Chapter 18—Using and Servicing Products and Structures

These chapters have been grouped into four units of study. Each area has at least one activity associated with it. The first unit includes Chapters 13-15 and focuses on locating, extracting resources, and converting them into industrial materials. During this unit the students will make paper or plywood.

The second unit focuses on manufacturing activities that convert standard stock (industrial materials) into products. This activity is supported with an open-ended design problem and a more structured fabrication problem.

The third unit investigates the processes used in constructing structures. Again, the students may complete either a design problem or a fabrication problem.

The final unit of study presents the process that consumers use as they select, use, service, and dispose of products and structures. This short activity has the students prepare an owner's manual for a simple product. During the study, the students will engage in four hands-on/minds-on activities.

Presenting Section 5

Estimated length—7 weeks

Section 5 is designed to help students understand technology as is applied to:

- Locating and extracting material resources
- Converting raw materials into industrial materials (standard stock)
- Manufacturing products from standard stock
- Constructing structures from standard stock (i.e., lumber, plywood, and PVC pipe) and manufactured products (i.e., windows, doors, and lavatories)
- Selecting, operating, maintaining, and disposing of (recycling or scrapping) products and structures

The content of Section 5 lends itself to being introduced through lecture/discussion strategies and reinforced by laboratory activities.

The 35-day schedule, shown in Figure 5-5, has been developed as a suggestion for scheduling both the presentations and activities for this section.

During this period of time, you will notice that four units of study (major topics) are introduced through a series of presentations and discussions. In addition, the students use common manufacturing and construction tools to complete either a design or a fabrication problem. **Note**: It is important that the students use tools and machines SAFELY. Demonstrate appropriate safety procedures to your class prior to any laboratory activity.

To prepare to present this section, the teacher should:

1. Read Chapters 13 through 18 in *Technology.*
2. Prepare discussions on:
   a. Production and technology (Chapter 13).
   b. Obtaining material resources (Chapter 14).
   c. Producing industrial materials (Chapter 15).
   d. Manufacturing (secondary) processes (Chapter 16).
   Note: The discussion of secondary manufacturing may be divided into three separate presentations, which are given as they are needed during Unit 2. These presentations are (1) sizing and shaping materials through casting, forming, and machining processes, (2) changing internal properties through conditioning processes, and (3) assembling and finishing products.
   e. Constructing processes (Chapter 17).
   Note: This presentation may be divided into three separate presentations, which are given as they are needed during Unit 3. These

presentations are (1) preparing to build—site work, (2) erecting the structure, and (3) finishing the structure and site.

f. Selecting and using products and structures (Chapter 18).

Note: Be sure to stress proper product selection, use, and maintenance and the proper and safe disposal of obsolete products.

3. Select a laboratory activity for each unit of study.

   Unit 1—Use either of two activities included in the Teacher's Resource—making paper or making plywood.

   Unit 2—Use either Activity 5A or Activity 5B. Activity 5A is an open-ended problem-solving manufacturing activity. Activity 5B is a structured activity in which groups of students use manufacturing technology to produce a simple product.

   Unit 3—Use either Activity 5C or Activity 5D. Activity 5C is a construction activity that involves an open-ended problem-solving approach to design and fabricate a nontraditional structure. Activity 5D is a structured construction activity in which groups of students construct and test a wall section for its sound absorption.

   Unit 4—Use the activity that has students preparing an owner's manual for a product produced in the class or selected by the teacher.

4. Practice the selected laboratory activities.
5. Prepare teaching materials (transparencies or bulletin boards, for example) to support the content.
6. Choose the study questions and test that will be used for the section.

Presenting the Activities

Consider the following recommendations in presenting the activity you select for this section.

Section 5: Production Technology	
Day	Activity
Unit 1—Obtaining and Converting Resources	
1	Discussion: "Production and Technology" (Chapter 13) Introduce primary processing laboratory activity
2–3	Discussion: "Obtaining Resources" (Chapter 14) Students work on laboratory activity
4–7	Discussion: "Primary Processing" (Chapter 15) Primary Processing Film Students complete laboratory activity
Unit 2—Manufacturing Products	
8	Discussion: "Manufacturing Processes" (Chapter 16) Introduce first processing activity
9–19	Demonstrate manufacturing processes Students work on laboratory activity
Unit 3—Constructing Structures	
20	Discussion: "Construction Processes" (Chapter 17) Introduce construction laboratory activity
21–31	Demonstrate construction practices Students work on laboratory activity
Unit 4—Maintaining Products and Structures	
32–34	Discussion: "Maintaining Artifacts" (Chapter 18) Servicing activity
35	Summarize section

Figure 5-5

Unit 1 Activity—Producing Industrial Materials

Students should produce an industrial material from raw materials. This type of processing is difficult to show in a technology education laboratory. Activities such as smelting iron, converting petroleum to plastics, and refining aluminum are complex and hazardous processes.

Still, students can experience the production of a common industrial material by one or both of the following activities.

Producing Paper

Students will produce a sheet of paper from reconstituted pulp. To complete the activity, students need the following materials and supplies:

Paper towels

Toilet tissue or newspaper

Hand mixer

Ink roller (brayer)

16-ounce cup

Electric iron

Production screen

8" × 10" tray

Divide the class into groups of two to four students for this activity. Have each group produce a sheet of paper using the procedures in the *Student Activity Manual.* If newspaper is used, the paper the students make will be gray in color.

Producing Plywood

1. Provide each student with five pieces of 1/8" × 3 1/2" × 3 1/2" veneer cut from a construction grade 2 × 4.
2. Have the students:
   a. Apply aliphatic resin or polyvinyl glue to one side of four pieces of veneer.
   b. Lay up a plywood sheet from the veneer.
   c. Place a sheet of waxed paper on each side of the plywood.
   d. Place the plywood in a vise and clamp the sheet for at least 45 minutes.
   e. Remove the plywood from the vise and let it cure for 24 hours.
3. The students should:
   a. Trim the sheet to 3" × 3".
   b. Sand the faces.
4. Each student can produce a peg game from the sheet of plywood. Some typical peg and board games are included in Section 3 of this guide.

Unit 2 Activity—5A Design Problem—Manufacturing Technology

This activity is designed to have the students think about the operations needed to produce a product. This product can be one from *Technology* or another of the students' choosing. To complete the activity they should complete these steps:

1. Sketch the product they will make. This can be the one in the book, a modification of that product, or an entirely different product.
2. They should prepare an operation sheet (such as those found in the *Student Activity Manual*) listing the operations they think are needed to produce the product.
3. Through discussion with the teacher, students should develop safety procedures for each operation and enter them on the operation sheet. The teacher should demonstrate these procedures, showing proper technique.
4. Each student should produce a product carefully, keeping a notebook (log) of each operation and the quality of the results.
5. Each student should select three steps in producing the product and complete the process analysis sheet (found in the *Student Activity Manual*).
6. Finally, they should select one casting and molding, one forming, and one separating process they completed or

observed. Each of these should be analyzed using the appropriate process review sheet (found in the *Student Activity Manual*).

Unit 2 Activity—5B Fabrication Problem—Manufacturing Technology

This activity is a structured introduction to manufacturing processes. It guides the students through the operations that are needed to produce a simple product. In each step, the students are expected to observe, analyze, and report on the processes they use. To complete the activity:

1. Introduce the product to the students by explaining the operation sheet on page 357 of *Technology.*
2. Demonstrate the proper procedures to be used in the manufacture of the product.
3. Have each student produce a product, keeping a log of each operation and the quality of the results.
4. Have each student select three steps in producing the product and complete the process analysis sheet (found in the *Student Activity Manual*).
5. Have the students select one casting and molding, one forming, and one separating activity they completed or observed. Each of these should be analyzed using the appropriate process review sheet (found in the *Student Activity Manual*).

Unit 3 Activity—5C Design Problem—Construction Technology

This activity is designed to have the students use their problem-solving abilities to produce structure to meet a specific need. To complete the activity, they should work in groups of two to four. The students should:

1. Define and describe the problem and its limitations.
2. Sketch a structure that will fulfill the customer's need.
3. Construct a model of the structure.
4. Present their solution to the class.

Unit 3 Activity —5D Fabrication Problem—Construction Technology

This activity is a structured introduction to construction processes. During the activity the class will work in groups of two to three students to build and test a wall section. During construction of the wall section, students will learn about framing, terms, and tool use. To the complete the activity:

1. Demonstrate wall framing.
2. Divide the class into groups of two to three students.
3. Assign each group to one of the wall sections shown in *Technology.*
4. Have each group:
   a. Fabricate their wall section.
   b. Calculate the material cost for it.
   c. Test its sound insulation qualities using at least four different intensities of sound according to the procedure listed in their textbooks.
   d. Graph the results of their test.
   e. Present their test results and cost data to the class.
   f. Chart and compare the data for the three different wall sections.
5. Determine the best construction technique in terms of sound insulation.
6. Compare the sound insulation qualities with the material cost for each wall section to determine the type of wall that is the most cost-effective.

Unit 4—Product and Structure Use Activity

A common technological action is servicing. This activity will allow students to troubleshoot and repair a technological device—a flashlight. To conduct this activity:

1. Obtain a flashlight for each two students in the class.
2. Obtain an inexpensive VOM or continuity checker for each group.
3. Create a problem with the flashlight by using any one of the following actions:
   a. Placing a burned-out bulb in it.
   b. Placing a dead battery in it.
   c. Coating the switch contacts with fingernail polish.

4. Divide the class into groups of two.
5. Discuss troubleshooting as a systematic elimination of possible problems.
6. Distribute the following items to each group:
   a. "Faulty" flashlight.
   b. Tester.
7. Have the students:
   a. Read the "Service Manual" in the *Student Activity Manual.*
   b. Conduct the troubleshooting procedure.
   c. Complete the "Job Ticket" and the "Customer's Bill" in the *Student Activity Manual.*

Taking Your Technology Knowledge Home

This take-home activity allows students to apply their knowledge of manufacturing and design and build a product they can sell or give to a charity. Typical simple products would be toy cars and trucks, CD holders, and pencil holders, for example. To complete the activity, the students will:

1. Decide whether they want to make a product to sell or give away.
2. Identify a type of product on which they will work.
3. Describe the specific function of the product.
4. List the features the product will have.
5. Identify the constraints within which they must work.
6. Sketch several possible designs for the product.
7. Select and refine the best design.
8. Build a prototype of the product.
9. Develop a procedure for making several products.
10. Design a package for the product.

Adjustments for Exceptional Students

The laboratory activities are designed to be done quickly. The level of sophistication can accommodate students of all ability levels. Gifted students should have more creative answers and possibly produce a more complex solution. Lower achievers can still solve the problem but with solutions within their ability level. In all cases, the amount of structure and teacher guidance should be adjusted to meet students' needs. Guard against providing too much structure, however, because this action often thwarts creativity and is less appealing to gifted students.

Also, careful grouping of students will allow each student to bring his or her own special abilities to the group. Some students will have more artistic ability, others might have more dexterity, and others might have more abstract reasoning skills. Throughout the course, group work allows students to develop cooperative work habits and to contribute their own unique abilities to the team.

Evaluation Procedures

The understanding students have of technology and technological systems should be measured in several ways, including:

- Observation and critical listening during class presentations and discussions.
- Worksheets and quizzes covering the cognitive material presented.
- Open-ended questions about the learned material and how technology is developed.
- Assessment of individual participation and contributions to group work.
- Assessment of the quality and creativeness of the solutions to the assigned design/fabrication activities.

Chapter 13

Using Technology to Produce Artifacts

Objectives

After studying Chapter 13, the students will be able to:

- Identify the two major types of production activities.
- Differentiate between activities involved in resource processing, product manufacturing, and structure construction.
- Define servicing.
- Define repair.

Answers to Test Your Knowledge Questions, *Text Page 253*

1. True
2. True
3. Evaluate individually.
4. False
5. True
6. True
7. Construction
8. Evaluate individually.
9. Servicing is the scheduled adjustment, lubrication, or cleaning required to keep a product or structure operating properly.
10. Repair involves fixing a broken, damaged, or defective product or building.

Answers to the Chapter Quiz, Teacher's Resource

1. True
2. False
3. False
4. True
5. False
6. c. Secondary manufacturing systems
7. b. Consumer products
8. b. Secondary processing activity
9. d. Natural resource
10. a. Manufacturing
11. b. Construction
12. b. Construction
13. a. Manufacturing
14. Servicing is the scheduled adjustment, lubrication, or cleaning required to keep a product or structure operating properly. Repair involves fixing a broken, damaged, or defective product or building.
15. Manufacturing activities are designed to create products. Construction activities are designed to create structures.

Caption for Transparency Master (Binder/CD Only)

Transparency Master 13-1. Production Activities—The goals of production activities are to produce products or structures. Shown are three types of production activities: process, manufacturing, and construction.

Using Technology to Produce Artifacts

Name ______________________________ Period ____________

Date ______________________________ Score ____________

Chapter 13 Quiz

True/False: If a statement is true, write *True* in the blank. If a statement is false, write *False* in the blank.

_________ 1. Product manufacturing activities are also referred to as secondary manufacturing systems.

_________ 2. Industrial materials are outputs of secondary manufacturing systems.

_________ 3. Servicing involves diagnosing a problem in a product and replacing or repairing broken parts.

_________ 4. Roads and canals are examples of civil structures.

_________ 5. Both industrial products and consumer products are outputs of resource processing activities.

Multiple Choice: Choose the best answer and write the corresponding letter in the blank.

_____ 6. Industrial products are outputs of:
   a. primary manufacturing systems.
   b. construction systems.
   c. secondary manufacturing systems.
   d. None of the above.

_____ 7. Which of the following types of products are designed for the end user in the product cycle?
   a. Industrial materials
   b. Consumer products
   c. Industrial products
   d. None of the above

_____ 8. Applying force to form materials is an example of a:
   a. primary processing activity.
   b. secondary processing activity.
   c. construction activity.
   d. None of the above.

_____ 9. Limestone is an example of a:
   a. standard material.
   b. industrial material.
   c. product.
   d. natural resource.

(Continued)

Name ______________________________

Matching: Match the following terms with their definitions.

_____ 10. Makes products in a building.

_____ 11. Creates civil structures.

_____ 12. Builds structures where they will be used.

_____ 13. Includes primary and secondary processing.

a. Manufacturing

b. Construction

Short Answer: In the space provided, write a brief answer to each of the following questions.

14. Explain the difference between servicing and repair.

____

____

____

____

____

15. Explain the difference between manufacturing and construction.

____

____

____

____

____

Chapter 14
The Types of Material Resources and How They Are Obtained

Objectives

After studying Chapter 14, the students will be able to:

- List the types of natural resources that are used as inputs to production systems.
- Identify the types of genetic materials that are used in production systems.
- Identify the types of fossil fuel materials that are used in production systems.
- Identify the types of minerals that are used in production systems.
- Describe how genetic materials are obtained for use in production systems.
- Describe how fossil fuel materials are located and obtained for use in production systems.
- Describe how minerals are located and obtained for use in production systems.

Answers to Test Your Knowledge *Questions, Text Pages 268–269*

1. Minerals
2. True
3. True
4. Petroleum, natural gas, and coal
5. ***Minerals*** are naturally occurring inorganic materials with homogenous composition, usually a crystalline structure, and distinct physical properties. Minerals differ from fossils fuels in that they have inorganic origins and are composed of a single element or compound.
6. Evaluate individually.
7. The major challenge for people dealing with genetic resources is to harvest the plant or animal at the proper stage of growth.
8. Yarding
9. Locating and extracting the material
10. This technique uses shock waves such as those in an earthquake. A small explosive charge is detonated in a shallow hole. The shock waves from the explosion travel into the earth. When the waves hit a rock layer they reflect back to the surface. Seismographic equipment uses two listening posts to measure the shock waves that bounce off various rock layers. Measuring the time it takes the waves to go down to the layers and reflect back allows the geologist to construct a map of the rock formations.
11. If an area has never produced oil or gas it is called a potential field. Producing fields are called proven reserves.
12. True
13. Surface or open-pit mining and underground mining
14. Evaporation
15. Fluid mining uses two wells that extend into the mineral deposit. Hot water is pumped down one of the wells. The water dissolves the mineral and is forced up the other well.

Answers to the Chapter Quiz, Teacher's Resource

1. True
2. False
3. False
4. True
5. True
6. b. Natural gas
7. d. Anthracite coal
8. c. Seed-tree cutting
9. b. Yarding
10. d. Both a. and b.
11. c. Shaft mining
12. a. Fluid mining
13. a. Lignite
14. Natural gas, petroleum, and coal
15. Typical genetic materials used in production systems are grains (wheat, oats, barley, and corn, for example), vegetable fibers (wood, flax, and cotton, for example), and animals (meat, hides, and wool, for example).

Caption for Reproducible Master

Reproducible Master 14-1. Manufacturing—Manufacturing is changing the form of materials to produce industrial or consumer products.

Caption for Transparency Master (Binder/CD Only)

Transparency Master 14-1. Securing Natural Resources—Genetic resources can be replanted after harvesting. Humans cannot create exhaustible resources. New sources must be found after they are extracted.

The Types of Material Resources and How They Are Obtained

Name ________________________________ Period ______________

Date ________________________________ Score ______________

Chapter 14 Quiz

True/False: If a statement is true, write *True* in the blank. If a statement is false, write *False* in the blank.

_________ 1. Genetic materials are obtained during the normal life cycle of plants or animals.

_________ 2. All petroleum has the same composition.

_________ 3. Searching for petroleum using a seismographic study involves measuring the strength of magnetic forces.

_________ 4. A producing oil field is called a proven reserve.

_________ 5. Open-pit mines for extracting minerals are generally deeper than those used to obtain coal.

Multiple Choice: Choose the best answer and write the corresponding letter in the blank.

_____ 6. Which of the following is composed primarily of methane?
- a. Petroleum
- b. Natural gas
- c. Coal
- d. Both a. and b.

_____ 7. Which of the following has the highest carbon content?
- a. Lignite
- b. Peat
- c. Bituminous coal
- d. Anthracite coal

_____ 8. Which method of harvesting trees removes all trees, regardless of species, except three or four per acre?
- a. Selective cutting
- b. Clear cutting
- c. Seed-tree cutting
- d. None of the above

_____ 9. Gathering logs at a central location called a landing is referred to as:
- a. cruising.
- b. yarding.
- c. bucking.
- d. felling.

_____ 10. Which of the following is a type of underground mining?
- a. Shaft mining
- b. Drift mining
- c. Open-pit mining
- d. Both a. and b.

(Continued)

Name ____

_____ 11. Which of the following requires a vertical shaft?
- a. Slope mining
- b. Drift mining
- c. Shaft mining
- d. Surface mining

_____ 12. Which of the following methods of obtaining minerals is referred to as the Frasch process and uses hot water to dissolve the mineral?
- a. Fluid mining
- b. Drift mining
- c. Evaporation
- d. None of the above

_____ 13. Which type of coal can be described as soft and porous?
- a. Lignite
- b. Bituminous coal
- c. Anthracite coal
- d. None of the above

Short Answer: In the space provided, write a brief answer to each of the following questions.

14. List three common fossil fuels.

____

____

____

15. List the types of genetic materials that are used in production systems.

____

____

____

____

____

Manufacturing

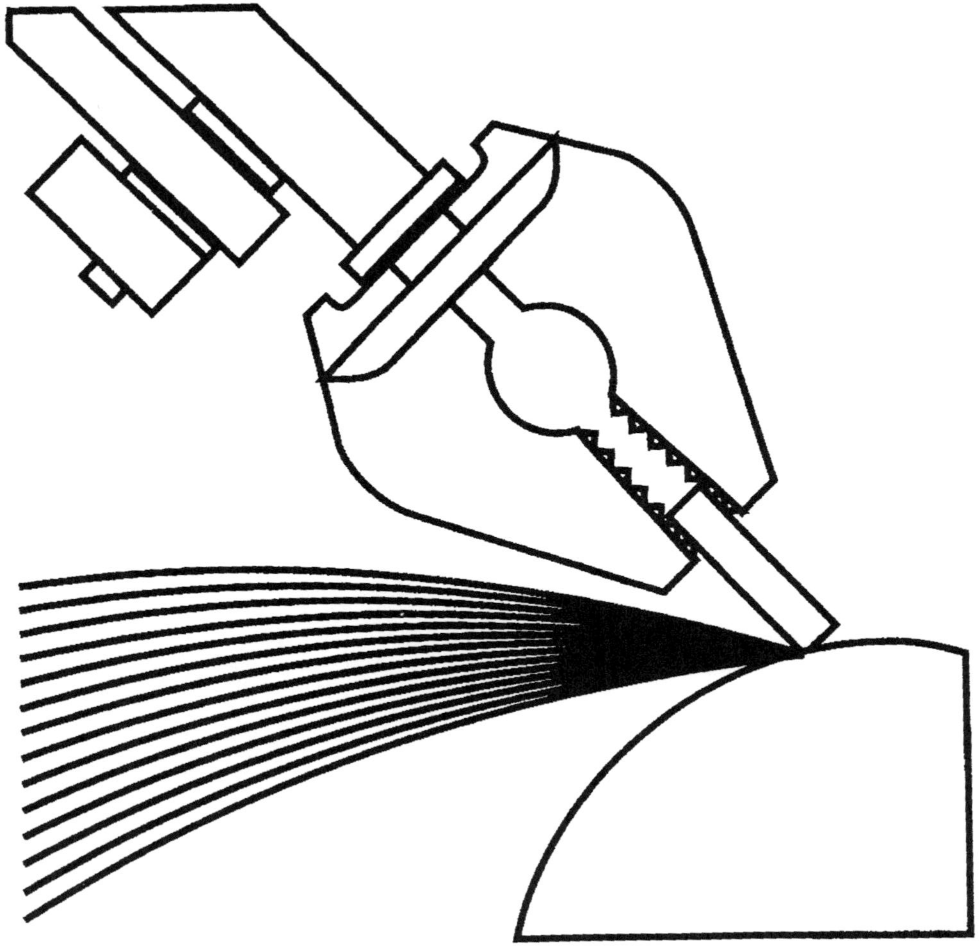

Changing the form of materials to produce industrial or consumer products

Chapter 15
Processing Resources

Objectives

After studying Chapter 15, the students will be able to:

- List the types of primary processes.
- Identify materials that result from mechanical processing.
- Describe characteristics of materials resulting from mechanical processing.
- Identify materials that result from thermal processing.
- Describe characteristics of materials resulting from thermal processing.
- Identify materials that result from chemical and electrochemical processing.
- Describe characteristics of materials resulting from chemical and electrochemical processing.

Answers to Test Your Knowledge *Questions, Text Pages 287–288*

1. Mechanical processes, thermal (heat) processes, and chemical and electrochemical processes
2. Mechanical
3. True
4. Veneer, lumber, and particleboard
5. Thermal and chemical processes
6. True
7. Blast
8. Galvanized
9. False
10. Float glass is formed by floating the molten glass on a bed of molten tin.
11. Separation, conversion, and treating
12. Fractional distillation breaks petroleum into major hydrocarbon groups.
13. Chemical and electrochemical processes
14. True
15. Pure aluminum oxide, the input to the second phase of making aluminum

Answers to the Chapter Quiz, Teacher's Resource

1. True
2. False
3. False
4. True
5. False
6. True
7. False
8. True
9. True
10. c. Chemical processing
11. b. Faces
12. a. Smelting
13. d. Both a. and c.
14. c. Electrochemical processing
15. b. Fractional distillation
16. a. Hardwood
17. b. Softwood
18. b. Softwood
19. a. Hardwood
20. b. Softwood

Caption for Reproducible Master

Reproducible Master 15-1. Types of Primary Processes—Primary processes use electrical/chemical, thermal (heat), and mechanical means.

Caption for Transparency Master (Binder/CD Only)

Transparency Master 15-1. Primary Processing—Primary manufacturing processes change natural (material) resources into industrial goods (standard stock).

Processing Resources

Name ______________________________ Period ____________

Date ______________________________ Score ____________

Chapter 15 Quiz

True/False: If a statement is true, write *True* in the blank. If a statement is false, write *False* in the blank.

_________ 1. The use of compression forces to crush a material is an example of a mechanical process.

_________ 2. Hardwood lumber is produced from needle-bearing trees.

_________ 3. An edger saw is used to debark logs.

_________ 4. Plywood resists warping better than solid lumber.

_________ 5. Tapping changes molten steel into sheets, plates, rods, and bars.

_________ 6. The process of cracking breaks heavier hydrocarbons into several smaller ones.

_________ 7. Galvanized steel is coated with tin to retard rusting.

_________ 8. Float glass is a sheet material used in windows.

_________ 9. Chemical and electrochemical processes are used to produce synthetic fibers, pharmaceuticals, and plastics.

Multiple Choice: Choose the best answer and write the corresponding letter in the blank.

_____ 10. Paper is produced through:
a. thermal processing.
b. mechanical processing.
c. chemical processing.
d. None of the above.

_____ 11. The outside layers of plywood are called:
a. crossbands.
b. faces.
c. cores.
d. None of the above.

_____ 12. Metals are extracted from ores through the process of:
a. smelting.
b. separation.
c. charging.
d. refining.

_____ 13. Pig iron, the basic input for steel, is produced through:
a. thermal processes.
b. electrical processes.
c. electrochemical processes.
d. Both a. and c.

(Continued)

Name ____

_____ 14. Aluminum is produced using:
a. mechanical processing.
b. thermal processing.
c. electrochemical processing.
d. Both a. and b.

_____ 15. Which of the following is a separation process?
a. Cracking
b. Fractional distillation
c. Polymerization
d. Both b. and c.

Matching: Match the following terms with their definitions.

_____ 16. Comes in random widths and lengths.

_____ 17. Comes from a needle-bearing tree.

_____ 18. Primarily used for construction.

_____ 19. Comes from trees that lose their leaves in the winter.

_____ 20. Comes in standard thicknesses, widths, and lengths.

a. Hardwood
b. Softwood

Types of Primary Processes

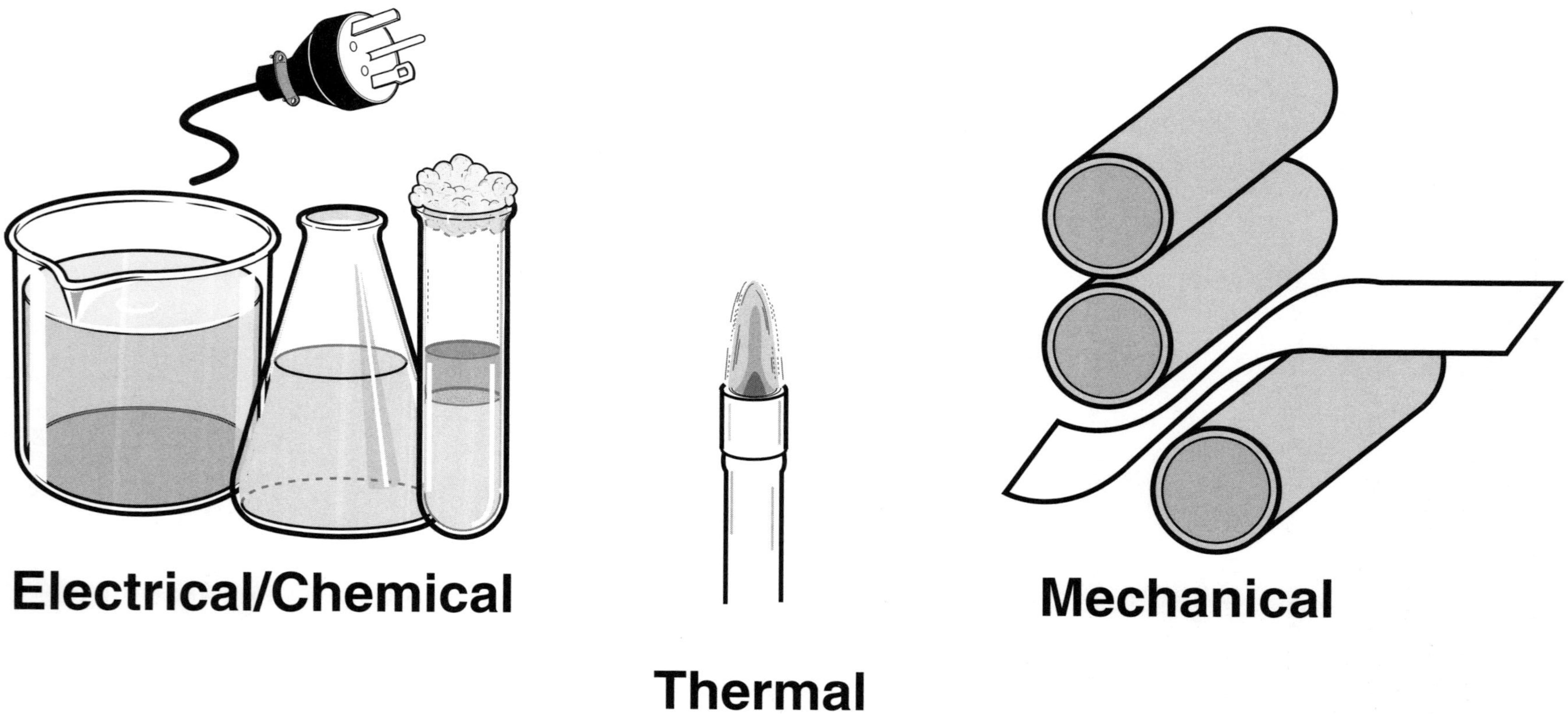

Chapter 16
Manufacturing Products

Objectives

After studying Chapter 16, the students will be able to:

- List the basic steps involved with casting and molding.
- Identify methods used in casting and molding processes.
- Name safety rules involved with casting processes.
- Identify the basic concepts of forming actions.
- Name the types of separating processes.
- Identify methods used in separating processes.
- Name safety rules involved with using machine tools.
- List the three types of conditioning processes.
- Identify methods used in conditioning processes.
- Identify methods used in assembling processes.
- Identify the methods used in finishing processes.
- Name safety rules involved in finishing processes.

Answers to Test Your Knowledge *Questions, Text Page 316*

1. Producing a mold of the proper size and shape. Preparing the material. Introducing the material into the mold. Solidifying the material. Extracting the product from the mold.
2. A mold that is destroyed to remove the cast item
3. Student answers may include any two of the following: Wear protective clothing and goggles when pouring a casting. Be sure everything to be used in the casting process is free of moisture. Place hot castings where they will not start a fire or burn someone. Keep the casting area orderly.
4. Cooling, drying, and chemical action
5. It is the pressure range between the yield point and the fracture point, which is the pressure range at which all forming processes operate.
6. The presence of a forming device, the application of force, and the consideration of material temperature
7. Mated dies are dies that have the desired shape machined into one or both halves of the die set. The material is placed between the die halves. The die set is closed, and the material is forced to take on the shape of the die cavities.
8. Evaluate individually. Common student answers include: pipe, tubing, and corrugated metal.
9. True
10. Shearing
11. A tool or other cutting element is always present. Motion occurs between the tool and work. The work and the tool are given support.
12. Reed
13. Student answers may include any of the following: Wear eye protection. Avoid loose-fitting clothing and keep long sleeves rolled up. Wear hearing protection when loud or high-pitched noises are present. Keep the laboratory clean. Wipe up any spills immediately. Wait until the machine has stopped to make adjustments, remove the workpiece, or clean up any scrap. Use a brush to clean chips off of a machine. Do not talk to anyone while running a machine. Most of all, if you do not understand how to run the machine, ask first.
14. False
15. Mechanical, chemical, and thermal conditioning
16. Student answers may include any two of the following: fusion bonding, flow bonding, pressure bonding, cold bonding, and adhesive bonding.
17. Student answers may include any of the following: staples, rivets, screws, nails, pins, bolts, and nuts. Other answers may be accepted at the instructor's discretion.

18. Student answers may include any two of the following: Wear eye and face protection. Apply finishes in a well-ventilated area; wear a respirator and protective clothing if solvents are toxic. Never apply a finish near open flame. Keep the finishing area orderly, and clean up any spills immediately. Store all rags and chemicals in approved containers. Use a spray booth (an enclosed area with fans to remove fumes) to remove toxic fumes. Dispose of any waste solvents and finishes properly; wash your hands to remove any finishes you have handled. Most of all, if you do not understand how to apply a finish, ask first.
19. True
20. Student answers may include any of the following: paint, varnish, enamel, and lacquer.

Answers to the Chapter Quiz, Teacher's Resource

1. False
2. True
3. True
4. False
5. False
6. False
7. True
8. d. Both a. and c.
9. b. Shell molding
10. c. Mated dies
11. a. Drawing machines
12. b. Milling machine
13. b. Annealing
14. c. Fusion bonding
15. Casting and molding processes—give materials shape by introducing a liquid material into a mold. Forming processes—apply force through a forming device to cause a material to change shape. Separating processes—remove excess material to make an object of the correct size and shape. Conditioning processes—change the internal properties of materials. Assembling processes—combine multiple parts to create a single product. Finishing processes—protect products and enhance their appearance.

Captions for Reproducible Masters

Reproducible Master 16-1. Casting and Molding—Shown are the five basic steps of the casting and molding process.

Reproducible Master 16-2. Forming—Shown are the three basic steps in the forming process. Forming processes are used to produce specific shapes in a material.

Reproducible Master 16-3. Separating—Shown are the three basic steps in the separating process. Separating processes remove excess material.

Reproducible Master 16-4. Conditioning—Shown are four steps of the conditioning process. Conditioning changes the internal properties of a material.

Reproducible Master 16-5. Assembly—Shown are four steps of the assembly process. Assemble means "to bring together."

Reproducible Master 16-6. Finishing—Shown are three steps in the finishing process. Finishing protects and beautifies a material.

Caption for Transparency Master (Binder/CD Only)

Transparency Master 16-1. Types of Manufacturing—Shows that natural resources are converted into industrial goods through primary processing and on into finished products through secondary processing.

Manufacturing Products

Name ______________________________ Period ______________

Date ______________________________ Score ______________

Chapter 16 Quiz

True/False: If a statement is true, write *True* in the blank. If a statement is false, write *False* in the blank.

_________ 1. Secondary manufacturing processes produce industrial materials.

_________ 2. Expendable molds are destroyed when the cast item is removed.

_________ 3. Injection molding produces many plastic products and uses permanent molds.

_________ 4. Hot forming a material builds internal stress in the material.

_________ 5. Nails are examples of semi-permanent fasteners.

_________ 6. Machining removes excess materials by fracturing the material.

_________ 7. Dipping is a method of applying finishing material that is widely used to produce galvanized products.

Multiple Choice: Choose the best answer and write the corresponding letter in the blank.

_____ 8. Which of the following gives size and shape to a material?
a. Casting and molding
b. Assembling
c. Separating
d. Both a. and c.

_____ 9. Which of the following casting processes uses a mold made of sand and resin mixture?
a. Investment casting
b. Shell molding
c. Green sand casting
d. None of the above

_____ 10. Drop forging produces the desired shape in a material by using:
a. open dies.
b. shaped dies.
c. mated dies.
d. None of the above.

_____ 11. Which of the following are used to push or pull materials through die openings?
a. Drawing machines
b. Hammers
c. Rolling machines
d. Both b. and c.

_____ 12. Which of the following uses a multiple-point, rotating cutter?
a. Turning machine
b. Milling machine
c. Drilling machine
d. Both a. and b.

(Continued)

Name ____

_____ 13. Which of the following softens and removes internal stress in a material?
a. Hardening
b. Annealing
c. Tempering
d. None of the above

_____ 14. Inert-gas welding is an example of:
a. pressure bonding.
b. flow bonding.
c. fusion bonding.
d. adhesive bonding.

Short Answer: In the space provided, write a brief answer to the following question.

15. List and briefly define the six types of secondary manufacturing processes.

____

____

____

____

____

____

Casting and Molding

Introducing a liquid material into a prepared cavity and causing the material to solidify.

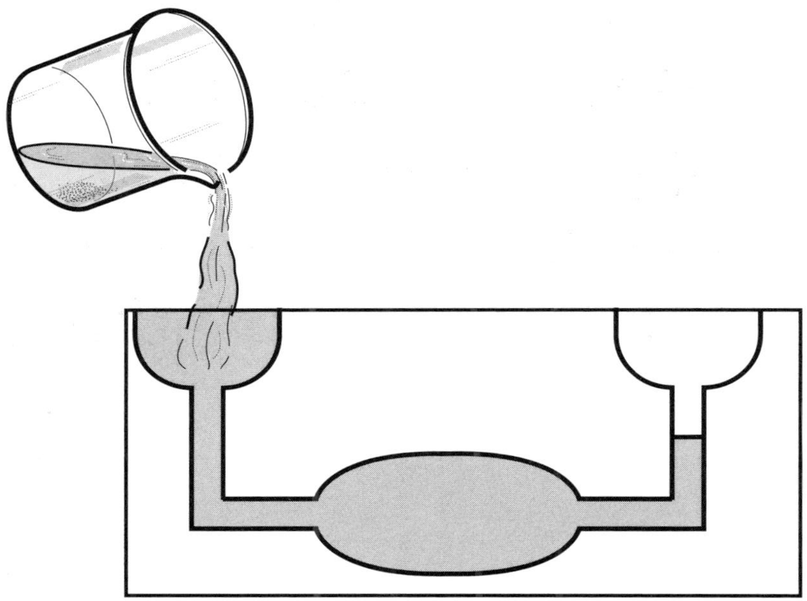

Prepare the Mold

Prepare the Material

Introduce the Material into the Mold

Cause the Material to Solidify

Extract the Product

Forming

Applying force through a shaping device to cause a material to change its shape permanently.

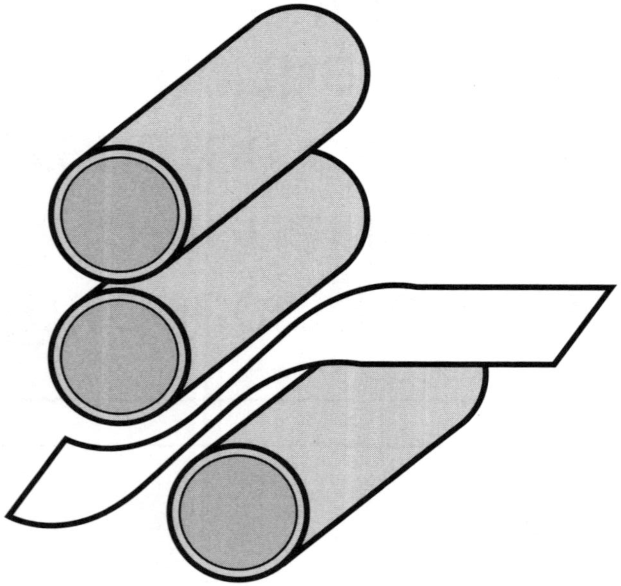

Develop the Shaping Device

Establish Material Temperature

Create the Forming Forces

Separating

Using a cutting element to remove excess material to create a desired size and shape.

Select the Cutting Device or Element

Support the Work and/or the Cutting Device

Create the Cutting and Feed Motions

Conditioning

Using heat, chemical action, or mechanical force to change the internal properties of a material.

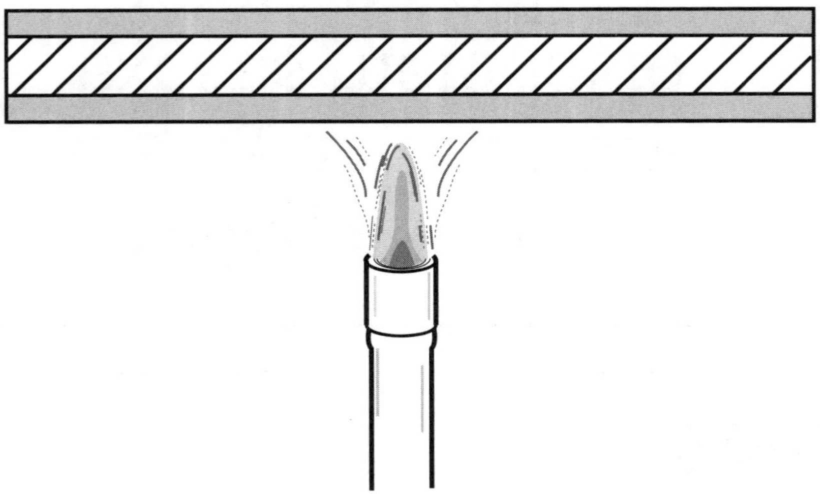

Determine the Desired Material Property

Determine the Material Structure Needed to Produce the Property

Select the Process to Develop the Property

Use Chemical, Thermal, or Mechanical Means to Condition the Material

Assembly

Temporarily, semi-permanently, or permantly attaching parts to make assemblies or products.

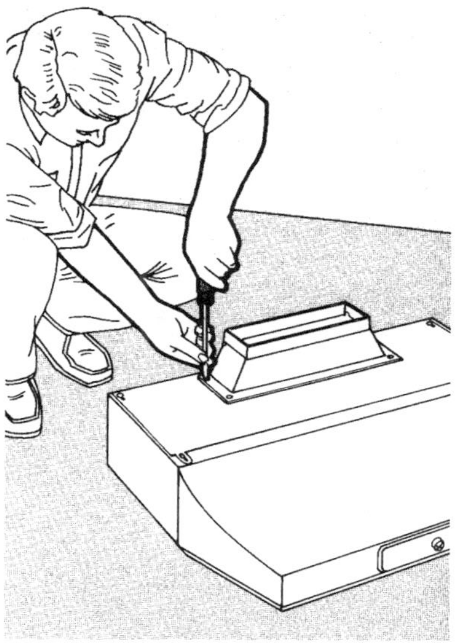

Determine the Type of Assembly Needed

Determine the Technique That Will Develop the Assembly

Select an Appropriate Joint for the Assembly

Install the Fastener or Bond the Material

Finishing

Coating or converting the surface of a material to protect and/or beautify it.

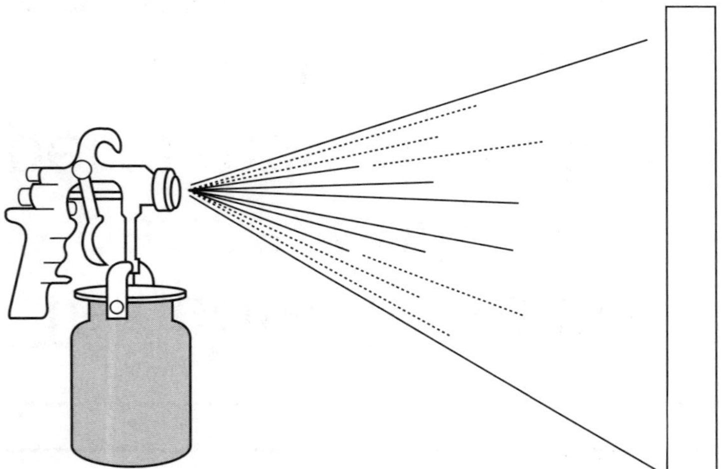

Select a Finishing Material or Technique

Prepare the Product for Finishing

Apply the Finish

Chapter 17
Constructing Structures

Objectives

After studying Chapter 17, the students will be able to:

- List the two types of construction.
- Describe the major types of buildings that are constructed.
- Identify the steps involved in constructing a structure.
- Identify the types of foundations.
- Describe the framework used in buildings.
- Describe how buildings are enclosed.
- Explain the types of utility systems used in buildings.
- Describe materials used in finishing buildings.
- Identify other types of buildings.
- Identify the characteristics of heavy engineering structures.

Answers to Test Your Knowledge *Questions, Text Page 340*

1. Buildings and heavy engineering structures
2. True
3. d. Enclosing the structure
4. f. Finishing interior and exterior
5. a. Preparing the site
6. b. Setting foundations
7. g. Completing the site
8. e. Installing utilities
9. c. Building the framework
10. c. Building the framework
11. b. Setting foundations
12. a. Preparing the site
13. c. Building the framework
14. f. Finishing interior and exterior
15. g. Completing the site
16. False
17. Lumber
18. A fascia board is used to finish the ends of the rafters and the overhang.
19. Potable water, wastewater
20. Fiberglass
21. Manufactured home
22. Elevation changes expressed in percentages
23. False
24. True
25. Arched

Answers to the Chapter Quiz, Teacher's Resource

1. False
2. False
3. False
4. True
5. False
6. True
7. True
8. b. Top plate
9. b. Fascia boards
10. b. Suspension bridges
11. d. The foundation
12. c. Header
13. b. Stud
14. e. Ceiling joist
15. a. Floor joist
16. c. Header
17. d. Rafter
18. b. Stud
19. Electrical, plumbing, climate control, and communications
20. Sheathing is the process of covering the exterior surfaces with an appropriate building material, such as plywood.

Captions for Reproducible Masters

Reproducible Master 17-1. Construction—Construction is erecting a structure on the site where it will be used.

Reproducible Master 17-2. Steps in Constructing a Structure—Listed are the seven basic steps used in constructing most structures.

Reproducible Master 17-3. Types of Civil Structures—Listed are a selection of important types of civil structures with some examples of each type.

Captions for Transparency Masters (Binder/CD Only)

Transparency Master 17-1. Types of Construction—Shown are two major classes of construction: building and heavy engineering.

Transparency Master 17-2. Types of Buildings—Buildings are grouped based on how they are used. Shown are three groupings of buildings: residential, commercial, and industrial.

Constructing Structures

Name ______________________________ Period ______________

Date ______________________________ Score ______________

Chapter 17 Quiz

True/False: If a statement is true, write *True* in the blank. If a statement is false, write *False* in the blank.

_________ 1. Spread foundations are used for buildings that are constructed on soft soil.

_________ 2. Ceiling joists rest only on exterior walls.

_________ 3. Most 110-volt circuits are limited to 220 amps.

_________ 4. Asphalt is generally applied in two layers when constructing roads.

_________ 5. Manufacturing facilities are an example of commercial structures.

_________ 6. When constructing a building, the structure is enclosed before utility systems are installed.

_________ 7. Soffits must have ventilation holes or vents.

Multiple Choice: Choose the best answer and write the corresponding letter in the blank.

_____ 8. Studs are tied together at the top with:
- a. sole plates.
- b. top plate.
- c. rafter plates.
- d. ceiling joists.

_____ 9. Which of the following is used to finish the ends of the rafters and overhang?
- a. The soffit
- b. Fascia boards
- c. Sheathing
- d. None of the above

_____ 10. Which of the following bridge types use cables to carry the loads?
- a. Cantilever bridges
- b. Suspension bridges
- c. Truss bridges
- d. None of the above

_____ 11. Which of the following is the most critical component of any structure?
- a. The framework
- b. The wall coverings
- c. The roof
- d. The foundation

(Continued)

Name ____

Matching: Match the following terms with their definitions.

_____ 12. A beam above a door.

_____ 13. A vertical member in a wall.

_____ 14. Weight-supporting member of a roof.

_____ 15. Weight-supporting member under a floor.

_____ 16. A beam above a window.

_____ 17. Angled board that rests on the top plate of the exterior walls.

_____ 18. Nailed to the top of the sole plate.

a. Floor joist
b. Stud
c. Header
d. Rafter
e. Ceiling joist

Short answer: In the space provided, write a brief answer to each of the following questions.

19. List the four major types of utilities.

____

____

____

____

20. What is sheathing?

____

____

____

____

____

Construction

Erecting a structure on the site
where it will be used.

Steps in Constructing a Structure

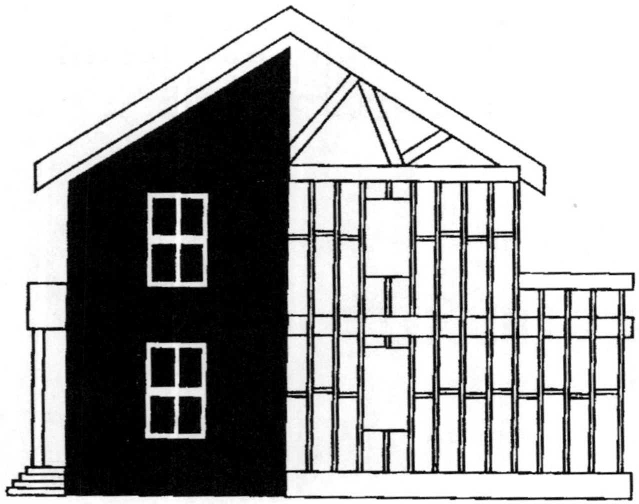

Preparing the Site

Setting Foundations

Building the Framework

Enclosing the Structure

Installing Utilities

Finishing the Interior and Exterior

Completing the Site

Types of Civil Structures

Transportation Structures
Roads, runways, rail lines, bridges, pipelines

Communication Structures
Broadcast towers, satellite dishes

Production Structures
Dams, power lines, substations

Chapter 18
Using and Servicing Products and Structures

Objectives

After studying Chapter 18, the students will be able to:

- Describe the factors to consider when selecting a product or structure.
- List the five steps used in installing a product.
- Describe factors involved in maintenance.
- List the steps used in repairing.
- Describe activities involved in altering products and structures.
- Explain the importance of recycling.
- Describe ways to dispose of materials.

Answers to Test Your Knowledge *Questions, Text Page 355*

1. Function, value, and appearance
2. Products are unpacked. Products are attached to utilities. Products are positioned and secured. Products are adjusted. Products are tested.
3. Preventive
4. The types of maintenance needed. Methods for performing maintenance. A time schedule for each maintenance task.
5. Diagnosis, replacement or adjustment, and testing
6. True
7. Evaluate individually.
8. Recycling can reduce the strain on both resources and landfill disposal sites.
9. False
10. Evaluate individually.

Answers to the Chapter Quiz, Teacher's Resource

1. True
2. False
3. True
4. False
5. True
6. False
7. False
8. False
9. True
10. True
11. c. Maintenance
12. b. Diagnosing, replacing, adjusting, and testing
13. c. Unpacking, positioning, adjusting, and testing
14. Function, value, and appearance
15. Needs are things that are necessary for living or performing essential tasks. Wants are requirements that are not necessary for living or performing essential tasks but are nice to have.

Caption for Reproducible Master

Reproducible Master 18-1. Steps in Using a Technological Product—Listed are the six steps involved in using a technological product properly.

Caption for Transparency Master (Binder/CD Only)

Transparency Master 18-1. Servicing—Shows the three steps in servicing: diagnosing, adjusting or replacing, and testing.

Using and Servicing Products and Structures

Name ______________________________ Period ____________

Date ______________________________ Score ____________

Chapter 18 Quiz

True/False: If a statement is true, write *True* in the blank. If a statement is false, write *False* in the blank.

_________ 1. Many food and yard wastes may be recycled in a compost bin.

_________ 2. Price and value mean the same thing.

_________ 3. Finding what is wrong with a malfunctioning product is called diagnosis.

_________ 4. Recycling is the sole responsibility of the product manufacturer.

_________ 5. In construction, altering means the same as remodeling.

_________ 6. All materials can be recycled.

_________ 7. Glass is the most common material put in landfills.

_________ 8. Maintenance is necessary only when a product malfunctions.

_________ 9. Adjustment is part of the installation process.

_________ 10. Nearly half of all aluminum is now coming from recycled beverage cans.

Multiple Choice: Choose the best answer and write the corresponding letter in the blank.

_____ 11. Scheduled tasks to keep a product working, such as oiling a machine, are called:
a. repairing.
b. diagnosis.
c. maintenance.
d. All of the above.

_____ 12. Repairing a product involves:
a. valuing, servicing, pricing, and adjusting.
b. diagnosing, replacing, adjusting, and testing.
c. unpacking, positioning, adjusting, and testing.
d. positioning, servicing, replacing, and testing.

_____ 13. Installing a product involves:
a. valuing, servicing, pricing, and adjusting.
b. diagnosing, replacing, adjusting, and testing.
c. unpacking, positioning, adjusting, and testing.
d. positioning, servicing, replacing, and testing.

(Continued)

Name ____ Period ____________________

Short Answer: In the space provided, write a brief answer to each of the following questions.

14. List the three things you should consider in selecting a product for purchase.

____

____

____

15. Explain the difference between needs and wants.

____

____

____

____

____

Steps in Using a Technological Product

1. Selecting
2. Installing
3. Maintaining
4. Repairing
5. Altering
6. Disposing

Section 6

Applying Technology: Communicating Information and Ideas

Teaching Materials

Text

Chapter 19, pp. 362–373
 Test Your Knowledge, pp. 372–373
 Applying Your Knowledge, p. 373

Chapter 20, pp. 374–399
 Test Your Knowledge, p. 394
 Applying Your Knowledge, p. 395

Chapter 21, pp. 400–417
 Test Your Knowledge, pp. 414–415
 Applying Your Knowledge, p. 415

Chapter 22, pp. 418–440
 Test Your Knowledge, pp. 436–437
 Applying Your Knowledge, p. 437

Chapter 23, pp. 442–461
 Test Your Knowledge, pp. 458–459
 Applying Your Knowledge, p. 459

Section Activities
 Activity 6A—Design Problem, p. 462
 Activity 6B—Production Problem, pp. 462–463

Student Activity Manual

Activity 6A—Design Problem, pp. 69–71
Activity 6B—Production Problem, pp. 73–75
Activity 6C—Telecommunication Activity, pp. 77–78
Taking Your Technology Knowledge Home, pp. 79–80

Teacher's Resources

Chapter 19 Quiz, pp. 202–203
Chapter 20 Quiz, pp. 209–210
Chapter 21 Quiz, pp. 218–219
Chapter 22 Quiz, pp. 222–223
Chapter 23 Quiz, pp. 226–227
Reproducible Masters
 19-1: Functions of Communication Media, p. 204
 19-2: Communication Process, p. 205
 19-3: Types of Communication Systems, p. 206
 20-1: Creating Media Messages, p. 211
 20-2: Designing the Message, p. 212
 20-3: Preparing to Produce the Message, p. 213
 20-4: Producing the Message, p. 214
 20-5: Delivering the Message, p. 215
 22-1: Telecommunication Systems, p. 224

Transparencies (Binder/CD only)
 19-1: Communication
 19-2: Impacts of Communication
 20-1: Ways to Communicate
 21-1: Types of Photographic Communication
 22-1: Developing and Delivering a Telecommunication Message
 23-1: Information and Communication Technology

Organization of Section 6

Section 6 is designed to present communication technology as a technological system that connects a sender with a receiver, Figure 6-1. Communication technology is different from language in that communication technology must have technical means involved.

The students will first explore the scope of communication technology. They will learn that five major types of communication technology

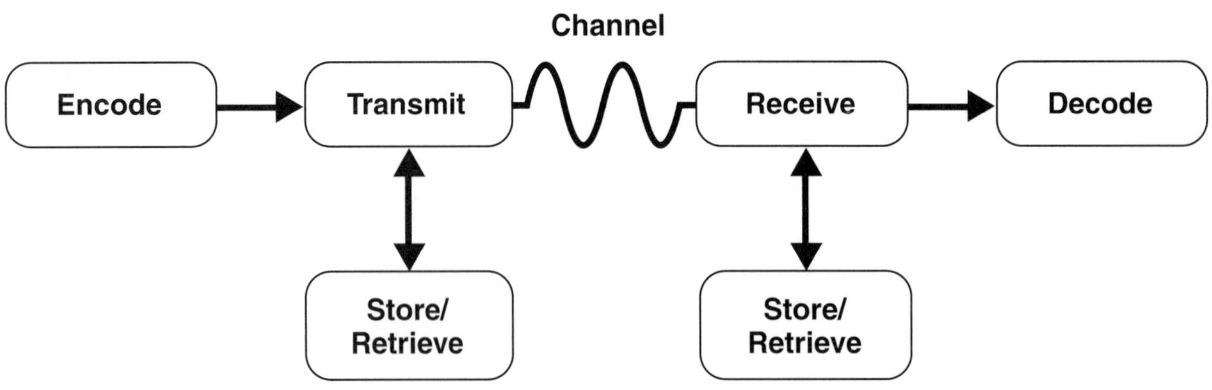

Figure 6-1

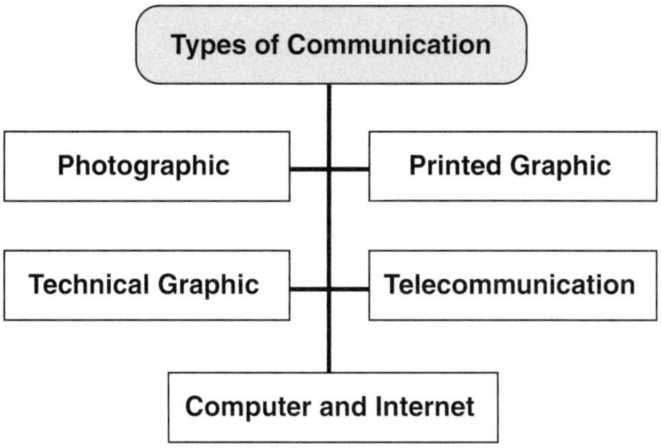

Figure 6-2

exist: printed graphic communication, photographic communication, telecommunication (electronic communication), technical graphic communications (which was presented in Section 4—Problem Solving and Design in Technology), and computer and Internet communication. See Figure 6-2.

Then they will address printed graphic, photographic, telecommunication, and computer and Internet communication independently. During the study of printed graphic communication they will learn about the six major types of printing, Figure 6-3:

- Relief: The oldest of all printing processes; prints an image that is raised above a surface
- Lithography: The most widely used method of printing; carries the image on a flat surface
- Gravure: Process in which the image carrier has the message chemically etched or scribed into its surface
- Screen: Process that uses a stencil with openings identical to the message
- Electrostatic: Process that uses a machine with a special drum and a strong light to create an electrically charged image that is transferred to paper
- Ink-jet: Process in which the message generated by a computer directs a special printer that sprays jets of ink onto paper to produce the message

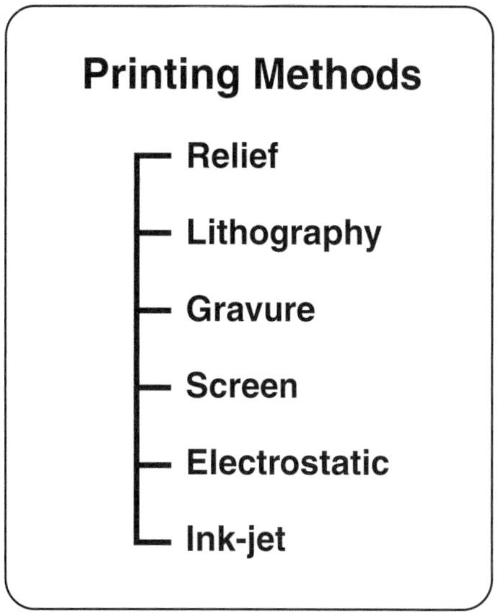

Figure 6-3

Following the introduction to the printing processes, the students will investigate the planning and production activities used to produce a printed product. These are:

- Designing the message
- Preparing to produce the message
- Producing the message

The second area of study will be photographic communication. In this area the students will learn about light waves, cameras, and film. They will then apply this knowledge as they design, produce, and reproduce photographic messages.

The third area of study in this section involves the students with telecommunication. They will explore two major types of telecommunication systems:

- Hard-wired systems: Systems that use electrical codes that are conducted over a permanent waveguide between the sender and the receiver
- Broadcast systems: Systems that use radio waves to carry the signal from the sender to the receiver

The students will use their knowledge of telecommunication systems as they complete two tasks: designing messages and producing and broadcasting messages. Specifically, they will complete the following steps under these two categories:

Designing Communication Messages

- Identifying the audience
- Selecting an approach
- Developing the message

Producing and Broadcasting Communication Messages

- Casting and rehearsing
- Performing and recording
- Broadcasting and receiving

During the final area of study in this section, the students will explore the newest area of communication: computer and Internet communication. They will review the parts of a computer system they studied in Chapter 7. They will then learn about computer networks and how they are united to form the World Wide Web and the Internet.

During this section, the students will study from five chapters and complete two laboratory activities. The chapters are:

- Chapter 19—Using Technology to Communicate
- Chapter 20—Printed Graphic Communication
- Chapter 21—Photographic Communication
- Chapter 22—Telecommunication
- Chapter 23—Computer and Internet Communication

Both activities should be used as the students complete this section. The design problem engages the students in printed graphic communications through a creative, problem-solving approach. The fabrication problem uses a more structured approach to studying photographic communication. In addition, two other activities that are included in this guide should be completed. These are:

- A telecommunication activity that includes the study of the printed graphic, photographic, and telecommunication systems
- An Internet activity

Presenting Section 6

Estimated length: 5 weeks

Section 6 is designed to help students understand communication technology as a means of using technical means (equipment and machines) to deliver information, feelings, and ideas. The students will explore communication technology through four basic media in this section:

- Printed graphic communication using printing processes
- Photographic communication using photography or digital imaging
- Telecommunication using electronically generated carrier signals
- Computer and Internet communication using computer systems to send and receive messages

The content of Section 6 should be delivered through presentations, discussions, and textbook readings. These approaches should be reinforced by at least two or three realistic laboratory activities.

The 25-day schedule, shown in Figure 6-4, has been developed for this section. During this period of time you will notice that five major topics are introduced through presentations and discussions. In addition, the students will receive equipment demonstrations and complete assigned laboratory activities.

To prepare to present this section, you should:

1. Read Chapters 19–23 in *Technology.*
2. Prepare discussions on:
    a. Using technology to communicate (Chapter 19)
    b. Printed graphic communication (Chapter 20)
    c. Photographic communication (Chapter 21)
    d. Telecommunication (Chapter 22)
    e. Computer and Internet communication (Chapter 23)

3. Prepare four laboratory activities.
   a. Activity 6A—Printed graphic communication—Design and produce a printed advertisement to promote technological literacy.
   b. Activity 6B—Photographic communication—Design and produce a picture set.
   c. Electronic communication—Develop and produce a 30-second commercial to promote recycling.
   d. Internet communication—Prepare a directory of Internet sites for a specific topic.
4. Practice the selected laboratory activities.
5. Prepare teaching materials (transparencies or bulletin boards, for example) to support the content.
6. Determine the study questions and quiz that will be used.

Presenting the Activities

Consider the following recommendations in presenting the activity you select for this section.

Activity 6A—Design Problem

This activity will introduce students to the work of the graphic designer and the producer of printed products. The students will be presented with a situation and challenge similar to the one below that is also in their textbook:

Situation:

You are employed as an advertising designer for Breckinridge and Rice Agency. One of your clients wants to promote technological literacy as a public service effort.

Challenge:

Design a full-page (6¾" × 9½") magazine advertisement that promotes technological literacy. It should encourage students to elect at least one class in technology education to help them understand technology as it impacts their lives. In this process: (1) develop a theme, (2) prepare a layout, and (3) specify type size and style, ink colors, and photograph type and location.

Section 6: Communication Technology	
Day	Activity
1	Discussion: "Using Technology to Communicate" (Chapter 19) Introduce laboratory activity 6A
2	Discussion: "Printed Graphic Communication (Chapter 20) Review sketching—rough and refined
3–4	Students work on laboratory activity—Planning phase
5–9	Demonstrate printing techniques Students work on laboratory activity—Production phase
10	Discussion: "Photographic Communication" (Chapter 21) Introduce laboratory activity 6B
11	Students work on laboratory activity—Planning phase
12–13	Demonstrate using a camera Students work on laboratory activity—Production phase
14	Demonstrate developing film and making a print Students work on laboratory activity
15	Discussion: "Telecommunications" Introduce telecommunications activity
16–17	Students work on laboratory activity—Planning
18–20	Demonstrate recording messages Students complete laboratory activity
21	Discussion: "Computers and the Internet" Students work on laboratory activity
22–24	Demonstrate Internet and World Wide Web Students complete laboratory activity
25	Summarize section

Figure 6-4

To complete the activity:

1. Divide the class into teams of four to five students.
2. Assign each team a theme for their advertisement such as promoting technological literacy, careful driving, or a drug-free school, for example.
3. Have each member of each team:
   a. Develop a slogan for their advertisement (for example, "Just Say No," "Buckle Up").
   b. Prepare three rough layouts for an advertisement using this theme.
   c. Select the best layout.
   d. Prepare a refined layout of the advertisement.
   e. Present the design to his or her group.
4. Have each group:
   a. Review the designs presented by its members.
   b. Select the best design.
5. Have each member prepare a modified comprehensive of the best design.
6. If time permits, have each group reproduce their designs using xerography, screen printing, or offset printing techniques.

Activity 6B—Production Problem

This activity is a fairly structured one in which the students first use storyboards to design a photographic message and then reproduce the message by using camera and darkroom techniques or digital imaging production and editing techniques. They will apply this knowledge as they meet the following challenge or one similar to it:

You are a communication designer for a publisher of children's instructional books. The company is developing a series of simple how-to-do-it booklets for children who are four to eight years old. Select a task that a child of this age must master. Design and produce a six-picture set that shows the child how to systematically complete the task.

To complete this activity they should follow the procedure listed in their textbook. At the appropriate time the students should receive informative demonstrations on:

1. Using a camera to expose film or produce a digital image.
2. Developing film in a darkroom or editing the digital image using computer software.
3. Making photographic prints or producing a set of ink-jet images.

Activity 6C—Telecommunication Activity

This activity is presented in this guide as the third communication activity. The students are challenged to produce a 30-second radio or television commercial to promote an issue of societal concern (for example, environmental protection, pollution, drug use, recycling).

This activity uses teams of three to five students to complete two major tasks, program development and program production. During the program development phase:

1. Assign each team a topic for their advertisement. (The same one can be assigned to each team, or each team can develop a separate one.)
2. Have each team member:
   a. Develop three themes for the assigned topic.
   b. Present the themes to his/her group.
3. Have each team select an appropriate theme.
4. Have each team member:
   a. Write a script and storyboard (for TV commercial) for the program.
   b. Present it to the team.
5. Have each team select the best script.

For the production phase of the commercial, each team should:

1. Develop production directions for camera location (for TV) and sound effects, for example.
2. Assign actor roles and production duties (recording engineer and camera operator, for example) to members of their group.
3. Rehearse the production.
4. Record and edit the production.
5. Present their production to the class.

Internet Activity

This activity is presented in this guide as the fourth communication activity. The students are challenged to develop an annotated bibliography of Internet sites that relate to a specific topic.

This activity uses teams of three to four students to complete the assignment.

1. Have the teams select a topic from a list you provide. A good source of topics is elementary teachers who are teaching technology sections.
2. Have each member of each team:
   a. Use search engines to locate sites for the topic.
   b. Select five appropriate sites.
   c. Record the name and URL address for the site on the laboratory sheet.
   d. Write a brief description of the site on the laboratory sheet.
3. Have each team select the best sites and prepare a master bibliography.

Reinforcing and Enriching

The following activities can be used to supplement and enrich the laboratory activities at the end of each textbook section and the Applying Your Knowledge activities included at the end of each textbook chapter.

1. Have all assignments focus on the same theme, such as stopping littering. Then the class should integrate these materials and present them as a package to such people as the school administration or city officials, for example.
2. During the telecommunication activity, have some groups produce radio commercials while other groups produce television commercials on the same theme. Have the class judge the effectiveness of these two media in communicating information and ideas.

Taking Your Technology Knowledge Home

This take-home activity allows students to apply their knowledge of communication systems to develop a sign for use in their neighborhood. Challenge the students to:

1. Work with their parents, guardians, neighbors, or youth group leader to identify a poster that could be used to promote an event such as a meeting or a garage sale, for example.
2. Establish the purpose for the sign.
3. Identify the information it will communicate.
4. Develop layout for the sign.
5. Present the layout to the proper adult for approval.
6. Produce the sign.
7. Give the sign to the appropriate adult.

Adjustments for Exceptional Students

The laboratory activities in this section are designed to be done quickly. Individual differences among students in a class can be accommodated by accepting varying levels of sophistication in the material presented. Also, communication media use a wide range of talents in their development. Students with artistic abilities, intellectual talents, writing skills, and production capabilities can all find a place to contribute to group work and feel needed. For this reason, all the activities in this section are designed for group work.

Evaluation Procedures

The understanding that students have of technology and technological systems should be measured in several ways, including:

- Observation and critical listening during class presentations and discussions.
- Worksheets and quizzes covering the cognitive material presented.
- Open-ended questions about the learned material and how technology is developed.
- Assessment of individual participation and contributions to group work.
- Assessment of the quality and creativeness of the solutions to the assigned design/fabrication activities.

Chapter 19
Using Technology to Communicate

Objectives

After studying Chapter 19, the students will be able to:

- Define communication technology.
- Describe items that are communicated.
- Name the three goals of communication.
- Identify features of the communication model.
- List the types of communication.
- Identify the five types of communication technology systems.

Answers to Test Your Knowledge *Questions, Text Pages 372–373*

1. Communication technology is a system that uses technical means to transmit information or data from one place to another or from one person to another.
2. Ideas, information, and emotions
3. Inform, persuade, and entertain
4. True
5. Noise
6. Human-to-human, machine-to-human, human-to-machine, and machine-to-machine
7. a. Printed graphic
8. c. Telecommunication
9. b. Photographic
10. e. Internet
11. d. Technical graphic
12. b. Photographic
13. d. Technical graphic
14. e. Internet
15. c. Telecommunication

Answers to the Chapter Quiz, Teacher's Resource

1. True
2. True
3. True
4. False
5. False
6. True
7. b. Information
8. c. Infotainment
9. a. Machine-to-human communication
10. b. Internet communications
11. a. Persuasion
12. Data are unorganized facts, information is organized data, and knowledge is information applied to a task.
13. Communication is the conveying of ideas, exchanging of information, and expression of emotions. Communication technology is a system that uses technical means to transmit information or data from one place to another or from one person to another.
14. Human-to-human, human-to-machine, machine-to-machine, and machine-to-human
15. Information is organized data, and noise is unwanted sounds or signals that become mixed in with the desired information.

Captions for Reproducible Masters

Reproducible Master 19-1. Functions of Communication Media—Communication can be used to inform, entertain, or persuade.

Reproducible Master 19-2. Communication Process—Shown are the steps taken in a televised message from the studio to televisions everywhere.

Reproducible Master 19-3. Types of Communication Systems—Listed here are the five types of communication systems that we come in contact with every day.

Captions for Transparency Masters (Binder/CD Only)

Transparency Master 19-1. Communication/Some Basic Terms—Shown are three important terms in communication and their definitions.

Transparency Master 19-2. Impacts of Communication—Shown are three possible impacts of communication. Communication can inform, stimulate, or challenge.

Using Technology to Communicate

Name ______________________________ Period ____________

Date ______________________________ Score ____________

Chapter 19 Quiz

True/False: If a statement is true, write *True* in the blank. If a statement is false, write *False* in the blank.

_________ 1. Knowledge is information applied to a task.

_________ 2. After a message is sent to a receiver, it must be decoded.

_________ 3. Electromagnetic waves are an example of a communication channel.

_________ 4. People cannot communicate emotions, only information.

_________ 5. Messages that are designed to entertain and educate are often called communitainment.

_________ 6. Static on your radio is an example of interference.

Multiple Choice: Choose the best answer and write the corresponding letter in the blank.

_____ 7. Which of the following is the foundation for decision making?
a. Emotions
b. Information
c. Ideas
d. None of the above

_____ 8. Providing information in an entertaining way is called:
a. edutainment.
b. communitainment.
c. infotainment.
d. entertainformation.

_____ 9. Stopping at a red light is a result of:
a. machine-to-human communication.
b. human-to-human communication.
c. human-to-machine communication.
d. None of the above.

_____ 10. Which of the following types of communication use a computer network to communicate?
a. Telecommunications
b. Internet communications
c. Technical graphic communications
d. All of the above

_____ 11. Causing people to act in a certain way is done through:
a. persuasion.
b. information.
c. entertainment.
d. None of the above.

(Continued)

Name ____

Short Answer: In the space provided, write a brief answer to each of the following questions.

12. Define each of the following: data, information, and knowledge.

____
____
____
____
____

13. What is the difference between communication and communication technology?

____
____
____
____
____

14. List the four combinations of senders and receivers that are used in communication technology.

____
____
____
____

15. What is the difference between information and noise?

____
____
____
____
____

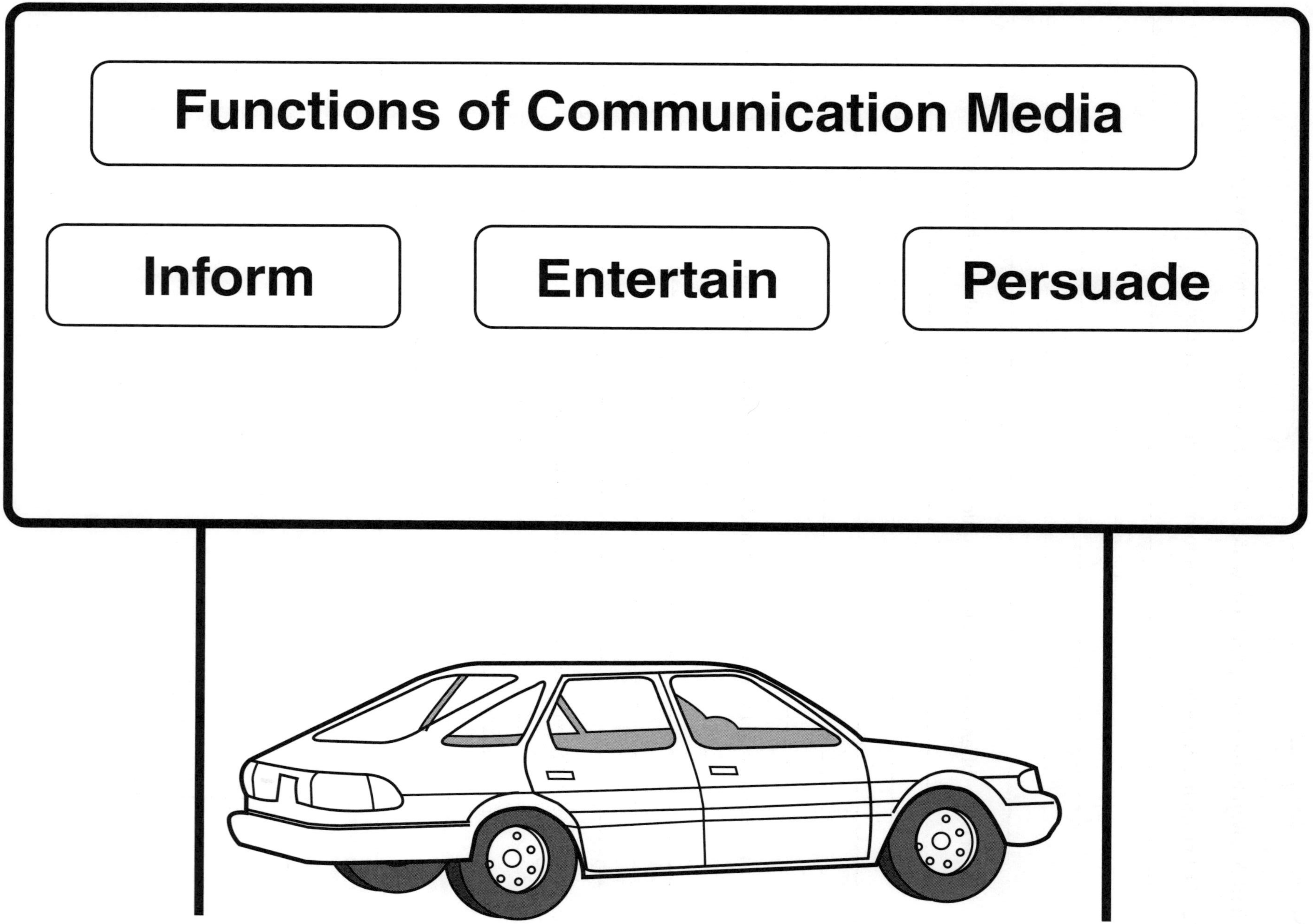
Functions of Communication Media
Inform
Entertain
Persuade

Communication Process

Encode - Transmit - Receive - Decode

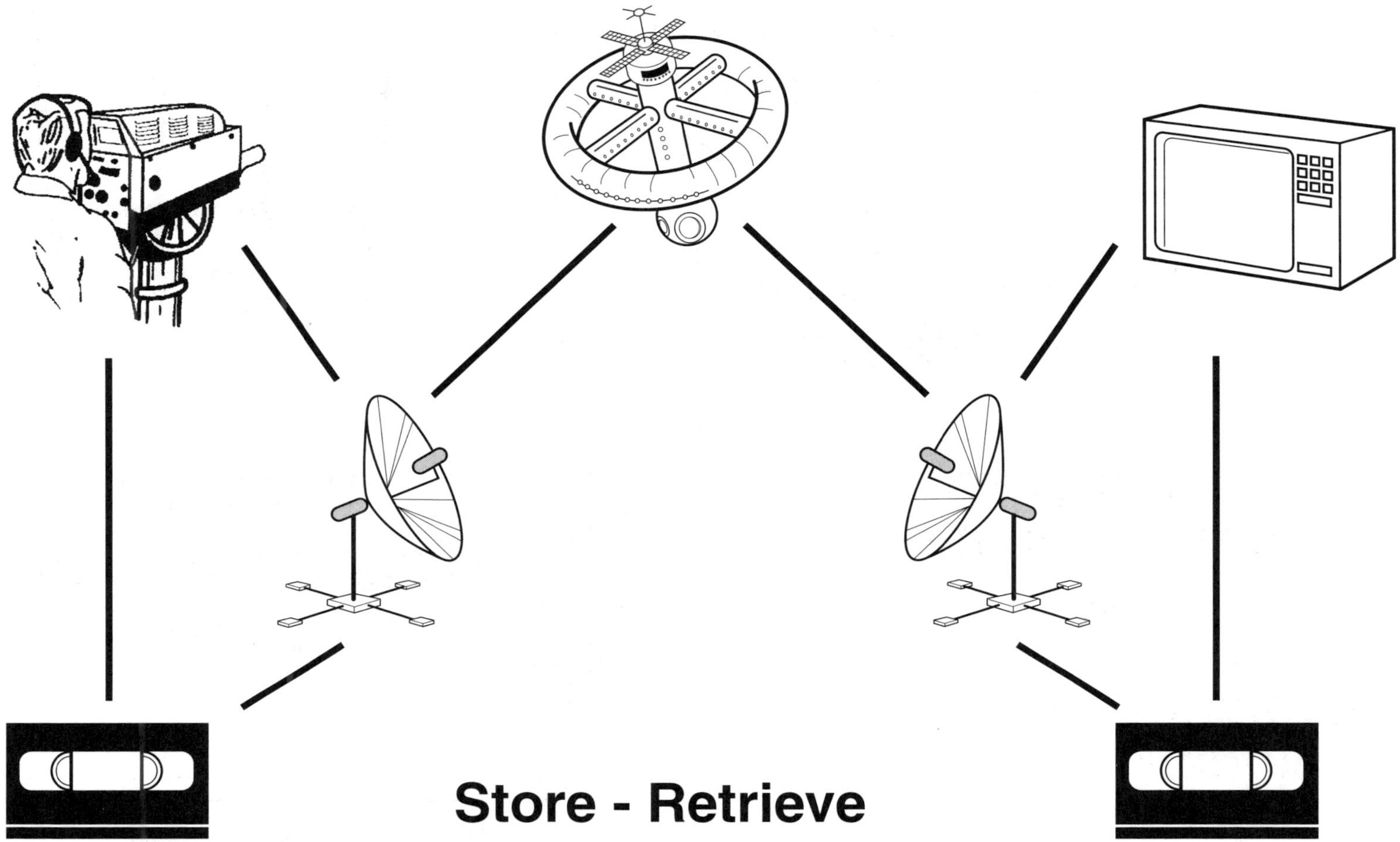

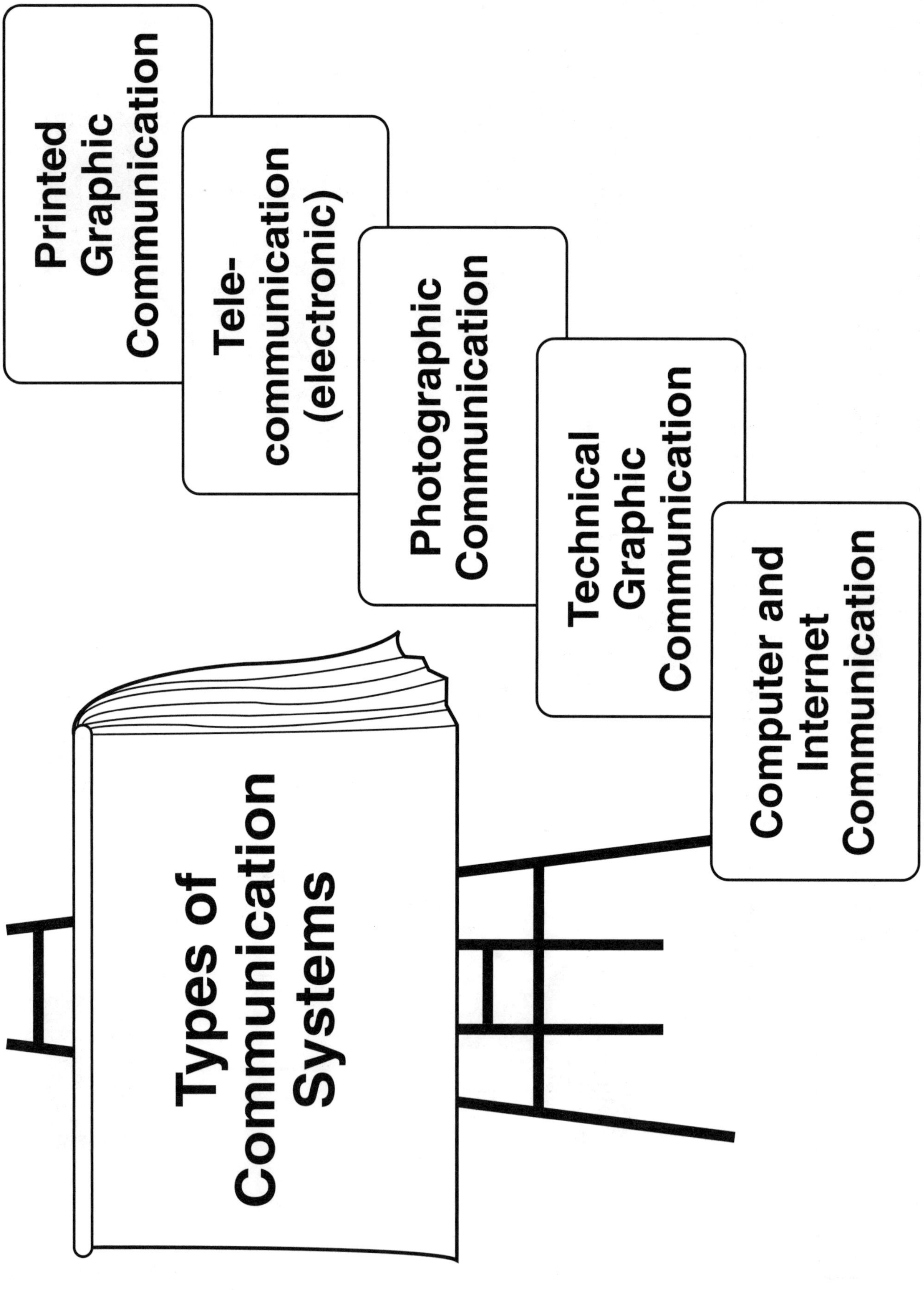
Types of Communication Systems
Printed Graphic Communication
Tele-communication (electronic)
Photographic Communication
Technical Graphic Communication
Computer and Internet Communication

Chapter 20
Printed Graphic Communication

Objectives

After studying Chapter 20, the students will be able to:

- Describe printed graphic communication.
- Name products produced by printed graphic communication processes.
- Identify the six major types of printing processes.
- List the steps involved in designing a printed graphic message.
- List principles of design.
- Identify how printed messages are prepared for production.
- Explain how printed graphic messages are produced.
- List activities involved with finishing printed graphic communication products.
- Describe computer-based publishing.

Answers to Test Your Knowledge *Questions, Text Page 394*

1. Printed graphic communication is the process of communicating information through symbols printed on a medium.
2. Student answers may include any three of the following: newspapers, magazines, books, brochures, pamphlets, labels, stickers, clothing designs, and signs. Additional answers may be accepted at the instructor's discretion.
3. d. Screen
4. e. Electrostatic
5. b. Lithographic
6. f. Ink-jet
7. b. Lithographic
8. c. Gravure
9. b. Lithographic
10. d. Screen
11. f. Ink-jet
12. c. Gravure
13. Designing the message. Preparing to produce the message. Producing the message.
14. Student answers may include any two of the following: proportion, balance, contrast, rhythm, variety, and harmony.
15. Layout
16. Composition, pasteup, and image carrier preparation
17. True
18. Evaluate individually.
19. Student answers may include any three of the following: cutting, die cutting, folding, drilling or punching, assembling, binding, stamping, and embossing.
20. The term desktop publishing is most often used to describe a fairly simple computer system that produces type and line illustration layouts for printed messages. The type and line illustrations are generated using two separate software packages. Electronic publishing employs more complex systems that can function as typesetting and layout systems. These systems produce and combine text and illustrations into one layout. Electronic publishing produces a higher quality product than desktop publishing produces.

Answers to the Chapter Quiz, Teacher's Resource

1. True
2. False
3. True
4. True
5. True
6. False
7. True
8. b. Lithography
9. d. Gravure printing
10. c. Proportion
11. b. Photochemical process
12. d. Registration unit
13. c. Binding
14. Relief, lithography, gravure, screen, electrostatic, and ink-jet
15. Designing the message. Preparing to produce the message. Producing the message.

Captions for Reproducible Masters

Reproducible Master 20-1. Creating Media Messages—Listed are four steps that take an idea and turn it into a finished media product.

Reproducible Master 20-2. Designing the Message—The design step of the process is the same for printed graphic media and telecommunication media.

Reproducible Master 20-3. Preparing to Produce the Message—Shown are the steps used in preparing to produce printed graphic media and telecommunication media. Both types of media involve considerable preparation.

Reproducible Master 20-4. Producing the Message—The third activity in creating a media message is producing the message. Shown are the production steps for printed graphic media and for telecommunication media.

Reproducible Master 20-5. Delivering the Message—The final step for a media message is delivery.

Caption for Transparency Master (Binder/CD Only)

Transparency Master 20-1. Ways to Communicate—Shown are four ways of communicating an idea—through speaking, writing, symbol, and picture.

Printed Graphic Communication

Name ______________________________ Period ______________

Date ______________________________ Score ______________

Chapter 20 Quiz

True/False: If a statement is true, write *True* in the blank. If a statement is false, write *False* in the blank.

_________ 1. Flexography is a type of relief printing that uses a plastic or rubber image carrier.

_________ 2. Office copiers use an ink-jet–printing process.

_________ 3. A page with informal balance has equal amounts of copy and illustration on each side of the layout.

_________ 4. Desktop publishing is used in many companies to produce newsletters, announcements, and flyers.

_________ 5. Color printing requires one pasteup for each color.

_________ 6. Gravure printing uses a raised surface like a rubber stamp.

_________ 7. Paper is an example of a printing substrate.

Multiple Choice: Choose the best answer and write the corresponding letter in the blank.

_____ 8. Today the most widely used method of printing is:
a. letterpress.
b. lithography.
c. gravure.
d. screen.

_____ 9. Stamps are printed with which of the following processes?
a. Offset lithography
b. Screen printing
c. Electrostatic printing
d. Gravure printing

_____ 10. The height-width relationship of a layout is called:
a. balance.
b. variety.
c. proportion.
d. contrast.

_____ 11. Most screen image carriers are produced using a(n):
a. photographic process.
b. photochemical process.
c. electromechanical process.
d. None of the above.

(Continued)

Name ____

_____ 12. Of the following parts in a press, which one positions the substrate for correct placement of the message?
a. Feeder unit
b. Printing unit
c. Delivery unit
d. Registration unit

_____ 13. Stapling pages together to form a magazine is referred to as:
a. assembling.
b. stamping.
c. binding.
d. die cutting.

Short Answer: In the space provided, write a brief answer to each of the following questions.

14. List the six common printing processes.

____

____

____

____

____

____

15. What are the three main steps in producing a printed graphic?

____

____

____

Creating Media Messages

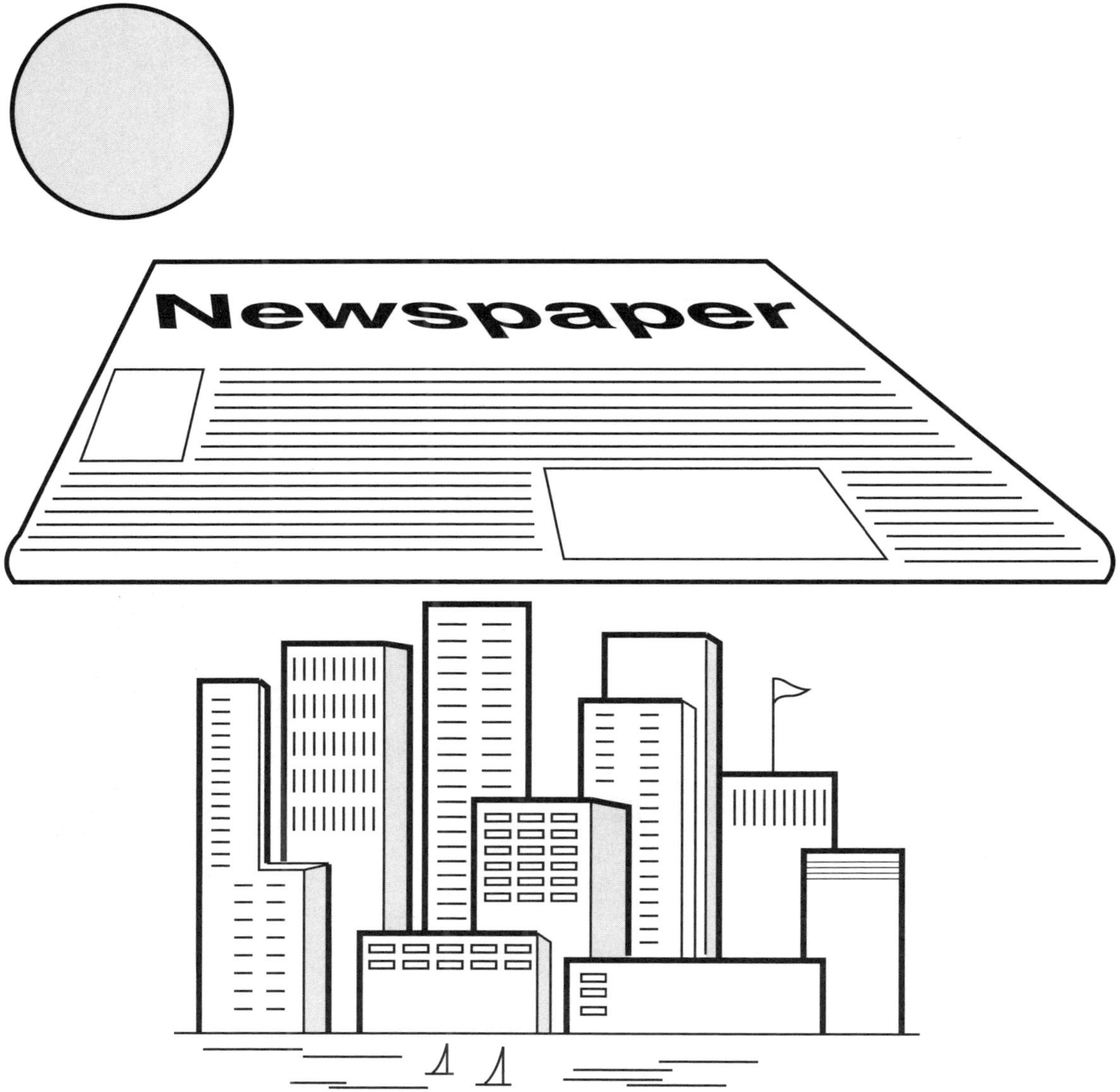

Designing the Message

Preparing to Produce the Message

Producing the Message

Delivering the Message

Designing the Message

Printed Graphic Media

Assess the Audience
Establish the Format
Gather Information
Write and Edit Copy
Design Graphics

Telecommunication Media

Assess the Audience
Establish the Format
Gather Information
Write and Edit Scripts
Design Graphics

Preparing to Produce the Message

Printed Graphic Media

Design the Layout
Generate the Copy
Generate Graphics
Prepare Pasteup

Telecommunication Media

Schedule Production
Cast
Stage
Rehearse

Producing the Message

Printed Graphic Media

Produce Image Carrier
Transfer Image
Finish Product

Telecommunication Media

Record Images
Edit Recording
Produce Master

Delivering the Message

Printed Graphic Media

Pack the Product
Ship to Customers
Read or Store
the Message

Telecommunication Media

Broadcast the Message
Receive the Message
View/Listen or Store
the Message

Chapter 21
Photographic Communication

Objectives

After studying Chapter 21, the students will be able to:

- Define photographic communication.
- List the three characteristics of light waves.
- List the three steps in developing a photographic communication message.
- Describe how photographic communication messages are designed.
- Differentiate between types of film.
- Explain the principles of a camera.
- Explain how photographic messages are reproduced.
- Differentiate between digital and traditional film photography.
- List common uses of transparencies.

Answers to Test Your Knowledge *Questions, Text Pages 414–415*

1. Photographic communication is the process of using photographs to convey an idea or information.
2. Amplitude, wavelength, and direction
3. Designing the message. Recording the image. Reproducing the message.
4. Student answers may include any of the following: Who is the intended audience? What is the message to be communicated? What type of photograph will best communicate the information or idea?
5. Point of interest
6. Student answers may include any of the following: using point of interest, using distance and scope, and using color.
7. Negative film produces a reverse image of the scene that is photographed. Positive transparencies produce the actual view, such that it can be viewed directly through a slide or movie projector.
8. Film exposing system, viewing system, film feeding system
9. d. Light meter
10. a. Shutter (**Note:** *b. Diaphragm* may also be accepted at instructor's discretion.)
11. b. Diaphragm
12. c. Lens
13. a. Shutter
14. c. Lens
15. b. Diaphragm
16. True
17. Developing, stopping, and fixing
18. Contact printing and projection printing
19. Evaluate individually.
20. Slides, filmstrips, and motion pictures

Answers to the Chapter Quiz, Teacher's Resource

1. False
2. True
3. False
4. True
5. False
6. True
7. False
8. True
9. c. Amplitude
10. d. All of the above
11. a. Diaphragm
12. c. The speed of the shutter should be increased and/or the aperture should be closed down.
13. b. Antihalation coating
14. Photography is the act of producing a photograph. Photographic communication is using these photographs to convey an idea or information.
15. Amplitude is the height of the wave. Wavelength is the distance from the beginning to the end of one wave cycle. Direction refers to the path that the light wave travels.

Caption for Transparency Master (Binder/CD Only)

Transparency Master 21-1. Types of Photographic Communication—Shown are still and motion types of film and digital photography.

Photographic Communication

Name ____ Period ____________________

Date ____ Score ____________________

Chapter 21 Quiz

True/False: If a statement is true, write *True* in the blank. If a statement is false, write *False* in the blank.

_________ 1. Panchromatic film is commonly referred to as color film.

_________ 2. Fast film is highly sensitive to light and is used to photograph objects in poorly lighted settings.

_________ 3. Stopping is a process that is used to make the image of a photograph permanent.

_________ 4. Contact prints are often used as proof prints.

_________ 5. Light waves must be transmitted by a material in order for us to see color.

_________ 6. Positive format film produces slides, which show the actual view photographed.

_________ 7. The longer the focal length of a lens, the smaller the objects will appear in the photograph.

_________ 8. The emulsion of color film is made up of three layers of light-sensitive material.

Multiple Choice: Choose the best answer and write the corresponding letter in the blank.

_____ 9. Which of the following indicates the intensity of a light wave?
a. Direction
b. Wavelength
c. Amplitude
d. None of the above

_____ 10. Electromagnetic waves include:
a. radio waves.
b. infrared light.
c. visible light.
d. All of the above.

_____ 11. Which of the following controls the amount of light that can enter a camera at any given moment?
a. Diaphragm
b. Shutter
c. Lens
d. All of the above

(Continued)

Name ____

_____ 12. When taking a picture in bright light:
- a. the speed of the shutter should be decreased and/or the aperture should be closed down.
- b. the speed of the shutter should be increased and/or the aperture should be opened.
- c. the speed of the shutter should be increased and/or the aperture should be closed down.
- d. the speed of the shutter should be decreased and/or the aperture should be opened.

_____ 13. Which of the following absorbs light and keeps it from reflecting back through the film?
- a. Silver halide crystals
- b. Antihalation coating
- c. Plastic base
- d. None of the above

Short Answer: In the space provided, write a brief answer to each of the following questions.

14. Explain the difference between photography and photographic communication.

____

____

____

____

____

15. List and explain the three properties of light waves.

____

____

____

Chapter 22
Telecommunication

Objectives

After studying Chapter 22, the students will be able to:

- Define telecommunication.
- List the major types of telecommunication systems.
- Explain the difference between frequency modulation and amplitude modulation.
- Differentiate between the two types of television stations.
- List the major activities involved in designing a broadcast message.
- Name activities involved in producing a broadcast message.
- Identify how broadcast messages are delivered.
- Describe a remote-link system.
- List the three mobile communication systems.

Answers to Test Your Knowledge *Questions, Text Pages 436–437*

1. Telecommunication means communicating over a distance.
2. Direct current
3. Hard-wired and broadcast
4. Evaluate individually.
5. UHF
6. Identifying the audience, selecting the approach, and developing the message
7. Record the sights and sounds of the performance
8. Reflected waves
9. A broadcast system in which the signal is generated at a remote location, transmitted to a relay antenna, and then relayed to the receiving antenna
10. Simplex system, duplex systems, and multiplex systems

Answers to the Chapter Quiz, Teacher's Resource

1. False
2. True
3. False
4. False
5. False
6. True
7. True
8. True
9. False
10. b. Transducers
11. c. Identifying the audience
12. d. Both a. and b.
13. b. Fact script
14. a. Magnetic tape
15. Simplex, duplex, and multiplex

Caption for Reproducible Master

Reproducible Master 22-1. Telecommunication Systems—There are two major types of telecommunications systems: hard-wired and broadcast.

Caption for Transparency Master (Binder/CD Only)

Transparency Master 22-1. Developing and Delivering a Telecommunication Message—Shows the steps in producing and delivering a telecommunication message.

Telecommunication

Name ______________________________ Period ______________

Date ______________________________ Score ______________

Chapter 22 Quiz

True/False: If a statement is true, write *True* in the blank. If a statement is false, write *False* in the blank.

_________ 1. Electrons flowing in one direction along a conductor is known as alternating current.

_________ 2. Radio signals move in all directions from the transmitting antenna.

_________ 3. Very high frequency (VHF) channels are usually assigned to public broadcasting and small independent television stations because their broadcast area is more limited than that of ultra high frequency (UHF) stations.

_________ 4. The director of a television show manages every element of the program.

_________ 5. Many newer communication systems convert digital signals into analog waves because digital signals are easily distorted.

_________ 6. The curvature of the earth limits the range of direct wave communication systems to approximately 60 miles.

_________ 7. The number of cycles per second is measured in hertz.

_________ 8. Multiplexing is frequently used to increase the capacity of a waveguide.

_________ 9. A remote-link broadcasting system allows a remote camera crew to broadcast directly to receiving antennas hundreds of miles away without relaying the signal.

Multiple Choice: Choose the best answer and write the corresponding letter in the blank.

_____ 10. Which of the following change one form of energy into another?
- a. Conductors
- b. Transducers
- c. Electrons
- d. None of the above

_____ 11. Which of the following is the first step in designing a television commercial?
- a. Selecting the approach
- b. Developing the message
- c. Identifying the audience
- d. None of the above

_____ 12. When selecting a communication approach, which of the following must be considered?
- a. The targeted audience
- b. The subject of the program
- c. The actors and producers
- d. Both a. and b.

(Continued)

Name ____

_____ 13. Which of the following types of scripts should be used when the performer is to ad-lib?
a. Show format script
b. Fact script
c. Detailed script
d. None of the above

_____ 14. Metal oxide crystals are arranged by magnetic forces in what type of recording medium?
a. Magnetic tape
b. Compact disc
c. Vinyl records
d. Both a. and b.

Short Answer: In the space provided, write a brief answer to the following question.

15. List the three mobile communication systems.

____

____

____

Telecommunication Systems

Hard-wired
Broadcast

Chapter 23

Computer and Internet Communication

Objectives

After studying Chapter 23, the students will be able to:

- List the parts of a computer system.
- Name the types of computer networks.
- List common uses of the Internet.
- Explain how computers are connected using the Internet.
- Define terms used in accessing the Internet.
- Identify the parts of domain names.
- Explain how web browsers and web servers work.
- Explain what hypertext markup language is.
- Explain what a hyperlink is.
- Define how search engines work.
- Explain how electronic mail works.

Answers to Test Your Knowledge *Questions, Text Pages 458–459*

1. Personal computer
2. Student answers may include any of the following: hard drives, floppy discs, CD-ROMs (compact disc—read-only memory), and DVDs. Additional answers may be accepted at the instructor's discretion.
3. Local area networks and wide area networks
4. Student answers may include any two of the following: the World Wide Web (WWW), electronic mail (e-mail), newsgroups or chat rooms, and electronic commerce (e-commerce).
5. True
6. Dial-up
7. An Internet protocol address is a number that identifies an individual computer connected to the Internet.
8. A uniform resource locator is a code that identifies a document or site on the World Wide Web.
9. True
10. Web browser
11. Hypertext markup language is a set of tags or codes that tell the receiving computer how a Web page should look.
12. Hyperlinks (or links)
13. A search engine is a special site that allows users to search for Web content by topic.
14. POP3
15. IP telephony

Answers to the Chapter Quiz, Teacher's Resource

1. False
2. True
3. True
4. False
5. False
6. True
7. True
8. True
9. False
10. False
11. c. Search engine
12. b. SMPT server
13. a. Hyperlink
14. d. IP telephony
15. The Internet Corporation for Assigned Names and Numbers (ICANN)

Caption for Transparency Master (Binder/CD Only)

Transparency Master 23-1. Information and Communication Technology—Shown is the gateway to the information highway.

Computer and Internet Communication

Name ______________________________ Period ______________

Date ______________________________ Score ______________

Chapter 23 Quiz

True/False: If a statement is true, write *True* in the blank. If a statement is false, write *False* in the blank.

_________ 1. Random-access memory can be read by the computer but cannot be changed.

_________ 2. A scanner is an example of an input device.

_________ 3. The largest WAN is the Internet.

_________ 4. A home computer connected to the Internet by a dial-up connection cannot be considered part of a network.

_________ 5. Because digital subscriber lines (DSL) use phone lines, a user must dial up the ISP every time he or she wishes to access the Internet.

_________ 6. The part of the URL before the colon indicates the format used to retrieve the document.

_________ 7. Three-segment domain names do not include the country code.

_________ 8. Each of the Web pages at a site will have its own URL.

_________ 9. HTML stands for homepage target metalinks.

_________ 10. One advantage of e-commerce is that the e-commerce companies do not need to advertise.

Multiple Choice: Choose the best answer and write the corresponding letter in the blank.

_____ 11. Google is an example of a(n):
a. domain.
b. newsgroup.
c. search engine.
d. Web browser.

_____ 12. Outgoing e-mail is handled by the:
a. domain name server.
b. SMPT server.
c. POP3 server.
d. Web server.

_____ 13. On a Web page, an underlined phrase that a user can click to connect to another page or site is known as a(n):
a. hyperlink.
b. router.
c. surf phrase.
d. search engine.

(Continued)

Name ____

_____ 14. A system that uses the Internet for voice messages is known as:
a. Talknet.
b. instant messaging.
c. chat rooms.
d. IP telephony.

Short Answer: In the space provided, write a brief answer to the following question.

15. Identify the group responsible for assigning domain names.

____

____

____

____

____

Section 7

Applying Technology: Transporting People and Cargo

Teaching Materials

Text

Chapter 24, pp. 466–475
- Test Your Knowledge, p. 474
- Applying Your Knowledge, p. 475

Chapter 25, pp. 476–505
- Test Your Knowledge, p. 504
- Applying Your Knowledge, p. 505

Chapter 26, pp. 506–521
- Test Your Knowledge, p. 518
- Applying Your Knowledge, p. 519

Section 7 Activities
- Activity 7A—Design Problem, p. 522
- Activity 7B—Fabrication Problem, pp. 522–523

Student Activity Manual

Activity 7A—Design Problem, pp. 83–84
Activity 7B—Fabrication Problem, pp. 85–87
Taking Your Technology Knowledge Home, pp. 89–91

Teacher's Resources

Chapter 24 Quiz, p. 236
Chapter 25 Quiz, pp. 241–242
Chapter 26 Quiz, pp. 250–251
Reproducible Masters
- 24-1: Transportation, p. 237
- 24-2: Transportation Systems, p. 238
- 25-1: Parts of a Transportation System, p. 243
- 25-2: Vehicles, p. 244
- 25-3: Vehicle Systems: Propulsion, p. 245
- 25-4: Vehicle Systems: Suspension p. 246
- 25-5: Vehicle Systems: Guidance, p. 247
- 25-6: Vehicle Systems: Structure, p. 248
- 26-1: Types of Transportation Systems, p. 252
- 26-2: Transportation Processes, p. 253

Transparencies (Binder/CD only)
- 24-1: Transportation Systems: Environment
- 24-2: Transportation Systems: Physical Elements
- 25-1: Vehicle Systems: Control
- 26-1: Transportation System

Organization of Section 7

This teaching section is designed to introduce students to transportation technology: its types, components, and vehicles, Figure 7-1. This study approaches transportation as systems that are used to move people and cargo from an origin to a destination. This action involves an environmental medium (land, water, air, and space), a vehicle that has five basic systems (structure, suspension, propulsion, guidance, and control), and support structures (terminals or pathways, for example).

- Transportation as technological action
- Transportation vehicles and their systems
- Operating transportation systems, including the integration of vehicles, pathways, and support structures into a dynamic system

During this study the students will first take a general look at transportation. They will study transportation from a systems point of view. Specifically, they will study:

- Land transportation: Highway systems, rail systems, and continuous flow systems
- Water transportation: Inland waterway and ocean-going systems
- Air transportation: Commercial aviation and general aviation
- Space transportation

This study will be followed by an exploration of the three components common to all transportation systems:

- Pathways or guideways
- Vehicles or carriers
- Support structures

Then the students will concentrate on the five systems that are commonly found in all transportation vehicles. These systems, shown in Figure 7-2, are structure, propulsion, suspension, guidance, and control.

Finally, the students will investigate the procedures used to operate a transportation system. They will look at both personal transportation systems (travel in a vehicle that is owned and operated by an individual) and commercial transportation systems (enterprises that move people and goods for money).

They will then learn that all commercial transportation systems have vehicles to contain and transport people and cargo, routes the vehicles

Transportation Technology		
Types of Transportation Systems	**Types of Transportation Systems Components**	**Transportation Vehicle Systems**
Land Water Air Space	Pathways Vehicles Support Structures	Structure Propulsion Suspension Guidance Control

Figure 7-1

travel from the origin to the destination, established schedules for the transportation of people and goods, and terminals at the origin and destination points of the system. While completing this section, students will read three chapters and complete one laboratory activity. The chapters are:

- Chapter 24—Using Technology to Transport
- Chapter 25—Transportation Vehicles
- Chapter 26—Operating Transportation Systems

Two different types of activities are provided for this section. If you want students to address the content from a problem-solving approach, select Activity 7A—Design Problem. If you want a more structured approach, select Activity 7B—Fabrication Problem.

Transportation Vehicle Systems	
Structure	The physical frame that protects cargo and people.
Propulsion	Moves the vehicle along its pathway.
Suspension	Supports the vehicle and its load as it moves along its path.
Control	Adjusts the speed and direction of the vehicle's travel.
Guidance	Provides information needed to direct the vehicle's travel.

Figure 7-2

Presenting Section 7

Estimated length: 3 weeks

Section 7 is designed to help students understand how technology is applied to transportation. They will learn that transportation:

- Is a system that uses technology.
- Generally uses vehicles to move people and cargo.
- Is a combination of vehicles along with support structures and activities.

The content of Section 7 is presented through two major media: the student textbook and lecture/discussion sessions. It is critically important to reinforce these presentations with realistic laboratory activities. The 15-day schedule shown in Figure 7-3 has been developed for this section. During this time, you will notice that three major topics are introduced through presentations and discussions. In addition, the students use simple tools to complete either a design or a fabrication problem. **Note**: It is important that the students use tools and machines SAFELY.

Section 7: Transportation Technology	
Day	Activity
1	Discussion: "Using Technology to Transport" (Chapter 24) Introduce laboratory activity
2–4	Demonstrate laboratory activity procedures Students work on laboratory activity
5–8	Discussion: "Transportation Vehicles" (Chapter 25) Students work on laboratory activity
9–14	Discussion: "Operating Transportation Systems" (Chapter 26) Students work on laboratory activity
15	Students present results of activity Summarize section

Figure 7-3

To prepare for presenting this section, you should:

1. Read Chapters 24, 25, and 26 in *Technology.*
2. Prepare discussions on:
   a. Using technology to transport people and cargo (Chapter 24).
   b. Transportation vehicles (Chapter 25).
   c. Operating transportation systems (Chapter 26).
3. Select a laboratory activity.
   a. Activity 7A—This is an open-ended problem-solving activity in which the students design and test a transportation vehicle kit.
   b. Activity 7B—This is a structured activity in which groups of students construct and test wind-powered vehicles. **Note:** Mousetrap or balloon-powered vehicles can be substituted for electric motors and propellers.
4. Practice the selected laboratory activity.
5. Prepare teaching materials (transparencies, or bulletin boards, for example) to support the content.
6. Determine which study questions and quiz questions will be used.

Presenting the Activities

Consider the following recommendations in presenting the activity you select for this section.

Activity 7A—Design Problem

This activity will introduce the students to transportation vehicles and their systems. The students will assume the role of an educational kit designer as they meet the following challenge:

You are employed as a designer for Technology Kits Company. This company markets simple kits that are fun to assemble and teach a basic principle of technology. Your most recent assignment is to design a transportation vehicle kit. Your supervisor wants a working model, a parts list for the kit, and a set of assembly directions for the purchaser.

To complete the activity:

1. Divide the class into design teams of two to three students.
2. Have each team:
   a. Select a type of vehicle (passenger, cargo, or a combination).
   b. An environmental medium (land, water, or air) for the vehicle.
3. Have each member of each team:
   a. Prepare three rough sketches of a vehicle that:
      (1) Is appropriate for its type and the environmental restrictions.
      (2) Can be packaged as an educational kit.
   b. Help select the best idea from the sketches.
   c. Prepare a refined sketch of the selected idea.
   d. Present the design to his or her group.
4. Have each group:
   a. Review the designs presented by its members.
   b. Select the best design.
   c. Develop a list of ways to improve the design.

5. Have each member prepare drafts of:
   a. A refined, detailed (dimensioned) sketch that incorporates the suggested improvements.
   b. A parts list for the kit.
   c. Directions for the customer to assemble the kit.
   d. Present the drafts to his/her group.
6. Have each group:
   a. Select the best drawing from among its members.
   b. Select the best parts list from among its members.
   c. Select the best assembly directions from among its members.
   d. Build a vehicle using the selected drawings, parts list, and directions.
   e. Present a completed model to the class.

Activity 7B—Fabrication Problem

This is a structured activity in which the students build and test a propeller-driven land vehicle from the drawings and directions in their textbook. They will apply their transportation knowledge as they meet the following challenges:

Build a basic land test vehicle that can be powered by a propeller. Develop an electric power system for the vehicle. Then, test the pitch of the propeller and the number of blades to determine its efficiency.

To conduct this activity make the following equipment available for each student or design team:

- One piece 1½" × 1½" × 11½" wood (pine, fir, spruce)
- Four wheels (available from many hobby stores)
- Two pieces ⅛" diameter × 3" welding rod (for use as the axles)
- Two pieces 1½" long of soda straw
- One low-voltage electric motor (available from a hobby or electronics store)
- One battery to match the motor voltage
- 12" electrical hook-up wire
- One each 4-blade and 3-blade propellers (available from a hobby store)

To complete this activity, the students should follow the procedure given in their textbook. At the appropriate time, the students should receive informative demonstrations on:

1. Fabricating the body
2. Installing the power (propulsion) system
3. Testing the vehicle and recording data

Reinforcing and Enriching

The following activities can be used to supplement and enrich the laboratory activities at the end of each textbook section and the Applying Your Knowledge activities included at the end of each textbook chapter.

1. Activity 7A—Design Problem can be enlarged to include a directional control system so the vehicle can be steered along a pathway.
2. Activity 7B—Fabrication Problem can be enlarged by having the students design auxiliary power systems using balloons, rubber bands, or mousetraps.

Taking Your Technology Knowledge Home

This take-home activity allows students to apply their knowledge of transportation systems to plan a trip. They will:

1. Work with their parents or guardians to establish a possible trip
2. Establish the purpose of the trip
3. Identify things to see or do along the way
4. Develop a route for the trip
5. Identify places to stay using Internet sites or motel/hotel guides
6. Estimate the cost of each day of travel
7. Summarize the trip and its costs

Adjustments for Exceptional Students

Both laboratory activities are designed to be done in a relatively short time—10 to 12 class periods. The level of expectation for the final product can be adjusted to accommodate students of different ability levels. Gifted students can be challenged to develop more creative answers and possibly produce a more complex solution. Lower achievers can still solve the problem but with solutions within their ability level.

With all students, the amount of structure and guidance should be adjusted to meet the students' needs. Avoid providing too much structure because this often stifles the creativity and has less appeal to gifted students.

Finally, careful grouping of students will allow each student to contribute special talents to the group. Some students may have more dexterity, others will have more artistic abilities; and still others may have greater abstract reasoning ability. Throughout this course, group work allows students to develop cooperative work habits while contributing unique abilities to the team.

Evaluation Procedures

The understanding that students have of technology and technological systems should be measured in several ways, including:

- Observation and critical listening during class presentations and discussions.
- Worksheets and quizzes covering the cognitive material presented.
- Open-ended questions about the learned material and how technology is developed.
- Assessment of individual participation and contributions to group work.
- Assessment of the quality and creativeness of the solutions to the assigned design/ fabrication activities.

Chapter 24
Using Technology to Transport

Objectives

After studying Chapter 24, the students will be able to:

- Define transportation.
- List types of transportation systems.
- Identify components of a transportation system.
- Define what is meant by place utility.
- Describe transportation pathways.
- Name types of transportation vehicles.
- Identify types of transportation support structures.

Answers to Test Your Knowledge *Questions, Text Page 474*

1. Transportation can be defined as all acts that relocate humans or their possessions.
2. False
3. Land, water, air, and space
4. The ability to move people or cargo from one place to another
5. True
6. Oceans
7. Evaluate individually.
8. Places where transportation activities begin and end
9. Intermodal shipping
10. True

Answers to the Chapter Quiz, Teacher's Resource

1. True
2. True
3. False
4. True
5. False
6. False
7. b. Intermodal transportation
8. a. Passenger
9. d. Space transportation
10. The ability to move people and cargo from place to place

Captions for Reproducible Masters

Reproducible Master 24-1. Transportation—Transportation involves relocating people or goods, often using a technical means.

Reproducible Master 24-2. Transportation Systems—Shown are two important elements that affect transportation systems: the physical elements and the environment.

Captions for Transparency Masters (Binder/CD Only)

Transparency Master 24-1. Transportation Systems/Environment—Transportation systems operate in many different environments.

Transparency Master 24-2. Transportation Systems/Physical Elements—Shown are the three major components of a human-made transportation system. Pathways link the elements of the system. Vehicles carry people and cargo along the pathway. Structures are used to gather, load, and unload passengers and goods.

Using Technology to Transport

Name ____ Period ____________________

Date ____ Score _____________________

Chapter 24 Quiz

True/False: If a statement is true, write *True* in the blank. If a statement is false, write *False* in the blank.

_________ 1. Transportation relocates people and goods.

_________ 2. Transportation systems have been developed for land, water, air, and space.

_________ 3. Rivers are part of the oceanic water transportation system.

_________ 4. Terminals are structures used to gather, load, and unload people and cargo.

_________ 5. Transportation vehicles carry people or cargo, but not both.

_________ 6. Pathways for all transportation systems are human-built structures.

Multiple Choice: Choose the best answer and write the corresponding letter in the blank.

_____ 7. Using more than one type of transportation system is called:
  a. merged transportation.
  b. intermodal transportation.
  c. bipolar transportation.
  d. polyvehiclular transportation.

_____ 8. Which of the following is not a necessary part of a complete air-transportation system?
  a. Passenger
  b. Support structures
  c. Vehicles
  d. Pathways

_____ 9. Which of the following can be considered an emerging technology?
  a. Land transportation
  b. Water transportation
  c. Air transportation
  d. Space transportation

Short Answer: In the space provided, write a brief answer to the following question.

10. Define *place utility.*

____

____

____

____

____

Transportation

Relocating people or goods

Transportation Systems

Physical Elements

Environment

Chapter 25
Transportation Vehicles

Objectives

After studying Chapter 25, the students will be able to:

- Identify the five systems present in a transportation vehicle.
- Identify the types of structural systems used in transportation vehicles.
- Name common propulsion systems used in transportation vehicles.
- Describe common suspension systems used in transportation vehicles.
- Identify common guidance and control systems used in transportation vehicles.

Answers to Test Your Knowledge *Questions, Text Page 504*

1. e. Control (**Note:** *d. Guidance* may also be accepted at the instructor's discretion.)
2. d. Guidance
3. a. Structure
4. c. Suspension
5. b. Propulsion
6. a. Structure
7. c. Suspension (**Note:** *a. Structure* may also be accepted at the instructor's discretion.)
8. e. Control (**Note:** *d. Guidance* may also be accepted at the instructor's discretion.)
9. Passenger or operator unit, cargo unit, and power unit
10. Internal combustion
11. True
12. Container ships
13. Buoyancy
14. True
15. A fuselage is the body of the aircraft, in which cargo and passengers are carried and to which the wings and tail assembly are attached.
16. Turbojet, turbofan, and turboprop
17. Thrust, lift, drag, and weight
18. Geosynchronous
19. Two solid fuel rockets strapped to the external fuel tank and the shuttle orbiter
20. False

Answers to the Chapter Quiz, Teacher's Resource

1. False
2. True
3. False
4. True
5. False
6. True
7. c. Suspension system
8. b. Submarine
9. a. Control system
10. a. Diesel engines
11. d. stern
12. b. Dry cargo ships
13. c. Hovercraft
14. a. Turbofan engine
15. The structural system provides spaces for people and cargo, power and control systems, and other devices. The propulsion system uses energy to produce power for motion. The suspension system supports the weight of the vehicle, passengers, and cargo as the vehicle moves along the pathway. The guidance system provides information about traffic conditions, location, speed, and direction of travel. The control system is used to change the speed and direction of travel of the vehicle.

Captions for Reproducible Masters

Reproducible Master 25-1. Parts of a Transportation System—Shown are two parts of the transportation system: the vehicle system and the support system.

Reproducible Master 25-2. Vehicles—Vehicle systems have five important subsystems listed here.

Reproducible Master 25-3. Vehicle Systems: Propulsion—Propulsion systems produce the force that powers a vehicle.

Reproducible Master 25-4. Vehicle Systems: Suspension—Suspension systems produce the proper support for a vehicle.

Reproducible Master 25-5. Vehicle Systems: Guidance—Guidance systems provide information needed to control a vehicle.

Reproducible Master 25-6. Vehicle Systems: Structure—Vehicle structure covers the important areas of the vehicle. This structure is based on the vehicle's use.

Caption for Transparency Master (Binder/CD Only)

Transparency Master 25-1. Vehicle Systems/Control—Control systems change the speed and direction of travel of a vehicle.

Transportation Vehicles

Name ______________________________ Period ____________

Date ______________________________ Score ____________

Chapter 25 Quiz

True/False: If a statement is true, write *True* in the blank. If a statement is false, write *False* in the blank.

_________ 1. Mechanical transmissions use valves to change hydraulic pressure so the transmission gearing can be changed.

_________ 2. A car's wheels and axles are part of its suspension system.

_________ 3. All ships are guided with a rudder.

_________ 4. Blimps are airships filled with helium.

_________ 5. An airplane's wings provide steering capability.

_________ 6. Satellites placed in geosynchronous orbit travel at the same speed the earth is turning.

Multiple Choice: Choose the best answer and write the corresponding letter in the blank.

_____ 7. Which of the following systems of a vehicle is designed to support the weight of the cargo?
a. Propulsion system
b. Control system
c. Suspension system
d. Both a. and c.

_____ 8. Which of the following vehicles has three degrees of freedom?
a. Train
b. Submarine
c. Automobile
d. None of the above

_____ 9. The steering wheel of an automobile is part of which system?
a. Control system
b. Guidance system
c. Propulsion system
d. Structure system

_____ 10. In which type of engine is only air drawn in during the first stroke of the four-stroke cycle?
a. Diesel engines
b. Gasoline engines
c. Both a. and b.
d. Neither a. nor b.

_____ 11. The back of a boat is called its:
a. bow.
b. starboard.
c. port.
d. stern.

(Continued)

Name ____

_____ 12. Which type of ship is used to transport vehicles?
a. Tankers
b. Dry cargo ships
c. Container ships
d. None of the above

_____ 13. Which of the following types of boats travel on a cushion of air between the hull and water?
a. Hydrofoil
b. Sailboat
c. Hovercraft
d. Submarine

_____ 14. Which type of engine is used in most commercial jets?
a. Turbofan engine
b. Turbojet engine
c. Turboprop engine
d. None of the above

Short Answer: In the space provided, write a brief answer to the following question.

15. List and describe the functions of the five vehicle systems.

____

____

____

____

____

Parts of a Transportation System

Support Systems

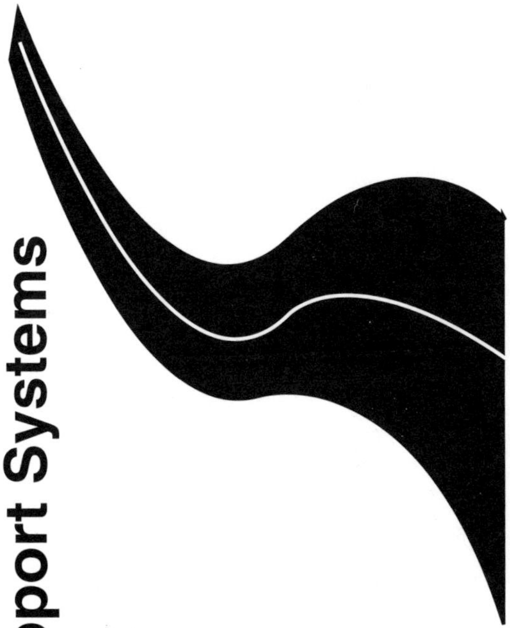

Vehicle Systems

Vehicles

Vehicles must contain, protect, and move people and cargo.

Vehicle Systems

Structure: Physical frame and covering

Propulsion: Converts energy into power to create movement

Suspension: Supports weight of the people and cargo

Guidance: Provides information needed to control the vehicle

Control: Changes speed and/or direction of the vehicle

Vehicle Systems

Propulsion

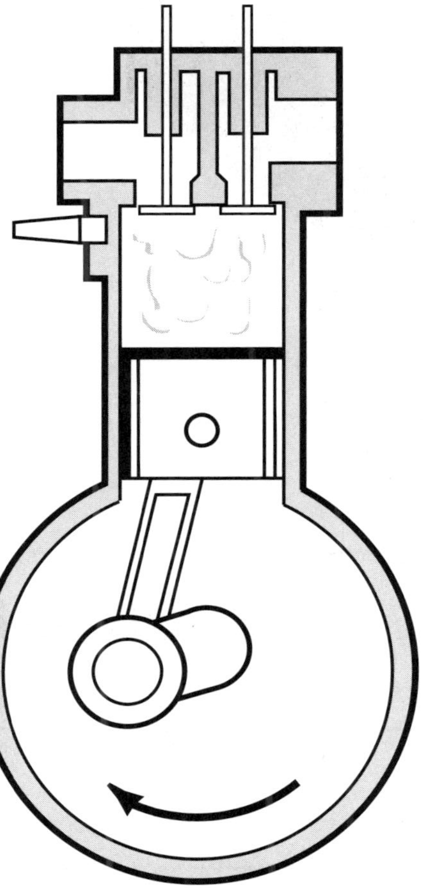

Converts energy into power to move the vehicle

Power generation

Power transmission

Vehicle Systems

Suspension

Supports the weight of the vehicle and the cargo

Mechanical (wheels, axles)

Fluids (air, oil, water)

Vehicle Systems

Guidance

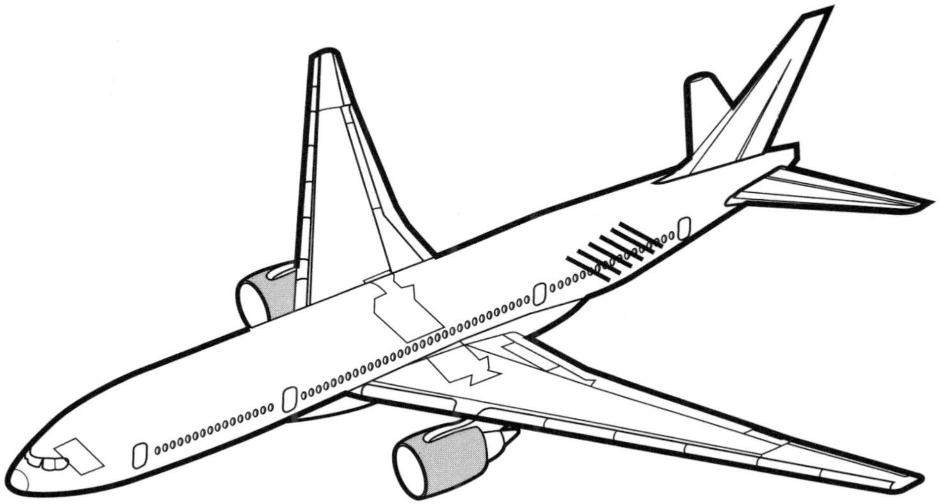

Provides information needed to control the vehicle's speed and direction

Vehicle location
Vehicle speed
Operating conditions
Rules and laws

Vehicle Systems

Structure

Provides the frame and covering for the major areas of the vehicle

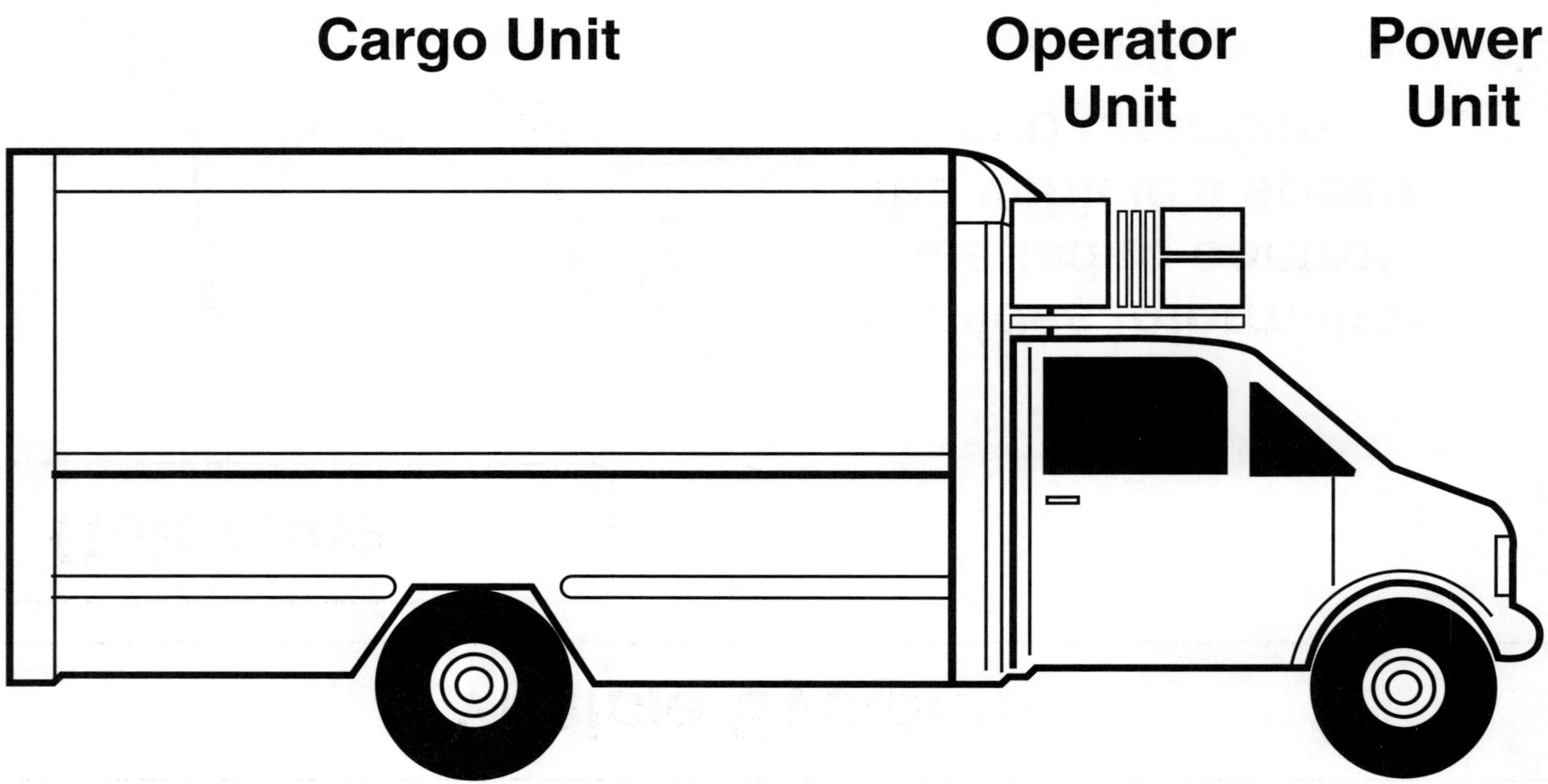

Chapter 26
Operating Transportation Systems

Objectives

After studying Chapter 26, the students will be able to:

- Describe how the speed of transportation has increased over time.
- Explain the differences between personal and commercial transportation systems.
- Identify the common components of all transportation systems.
- Name the major actions of transportation systems.
- Describe maintenance and repair activities involved in transportation systems.
- Identify various regulatory agencies that have an effect on transportation systems.

Answers to Test Your Knowledge *Questions, Text Page 518*

1. Evaluate individually.
2. Evaluate individually.
3. A vehicle, a route the vehicle travels from the origin to the destination, an established schedule for the movement of people and goods, and terminals at the origin and destination points of the system
4. A hub-and-spoke system is made up of small local terminals and large regional terminals. The local terminals are on the ends of two-way routes that radiate from the regional terminal. The routes appear to be spokes attached to a hub of a wheel.
5. False
6. Loading, moving, and unloading
7. Evaluate individually.
8. Evaluate individually.
9. Sets highway construction standards for federally financed roads and regulates the safety of interstate trucking
10. False

Answers to the Chapter Quiz, Teacher's Resource

1. False
2. True
3. False
4. False
5. True
6. True
7. False
8. True
9. b. fuel inefficiency
10. c. A highway connecting several cities
11. a. Loading
12. d. Federal Aviation Administration
13. Privately owned by individual or group of investors—profit-driven enterprises that use private funds to finance their initial operation. Government owned—provide transportation services where there is little chance for a private company to make a profit.
14. The four components of a commercial transportation system are vehicles, routes, established schedules, and terminals.
15. The three major actions are loading, moving, and unloading.

Captions for Reproducible Masters

Reproducible Master 26-1. Types of Transportation Systems—Shown are a few categories into which transportation systems can be broken.

Reproducible Master 26-2. Transportation Processes—Listed are five transportation processes that help move people and cargo.

Caption for Transparency Master (Binder/CD Only)

Transparency Master 26-1. Transportation System—Shows that routes, schedules, vehicles, and terminals make up a transportation system.

Operating Transportation Systems

Name ______________________________ Period ______________

Date ______________________________ Score ______________

Chapter 26 Quiz

True/False: If a statement is true, write *True* in the blank. If a statement is false, write *False* in the blank.

_________ 1. Most commercial transportation systems are government owned.

_________ 2. Passenger terminals are more complex than cargo terminals.

_________ 3. Only domestic transportation is regulated.

_________ 4. Repairing is defined as the routine tasks that keep a vehicle running properly.

_________ 5. Most personal transportation systems use public pathways.

_________ 6. Both trucking companies and airlines use hub-and-spoke systems to move cargo.

_________ 7. The Federal Highway Administration sets speed limits for highways.

_________ 8. Buses and rapid transit lines are often owned and operated by city or county governments.

Multiple Choice: Choose the best answer and write the corresponding letter in the blank.

_____ 9. The main disadvantage of personal transportation systems is:
- a. inflexible schedules.
- b. fuel inefficiency.
- c. government ownership.
- d. None of the above.

_____ 10. Which of the following is an example of an arterial road?
- a. A country road leading to farms
- b. A city street
- c. A highway connecting several cities
- d. An alley

_____ 11. Vehicles receive scheduled services, such as checking the oil, during which phase of transportation?
- a. Loading
- b. Moving
- c. Unloading
- d. Both a. and c.

_____ 12. Which of the following agencies regulates air space?
- a. Federal Maritime Commission
- b. Federal Highway Administration
- c. Interstate Commerce Commission
- d. Federal Aviation Administration

(Continued)

Name ______________________________

Short Answer: In the space provided, write a brief answer to each of the following questions.

13. List and describe the two types of ownership of commercial transportation systems.

____
____
____
____
____

14. What are the four components of a commercial transportation system?

____
____
____
____

15. List the three major actions that it takes to transport people and cargo.

____
____
____

Types of Transportation Systems

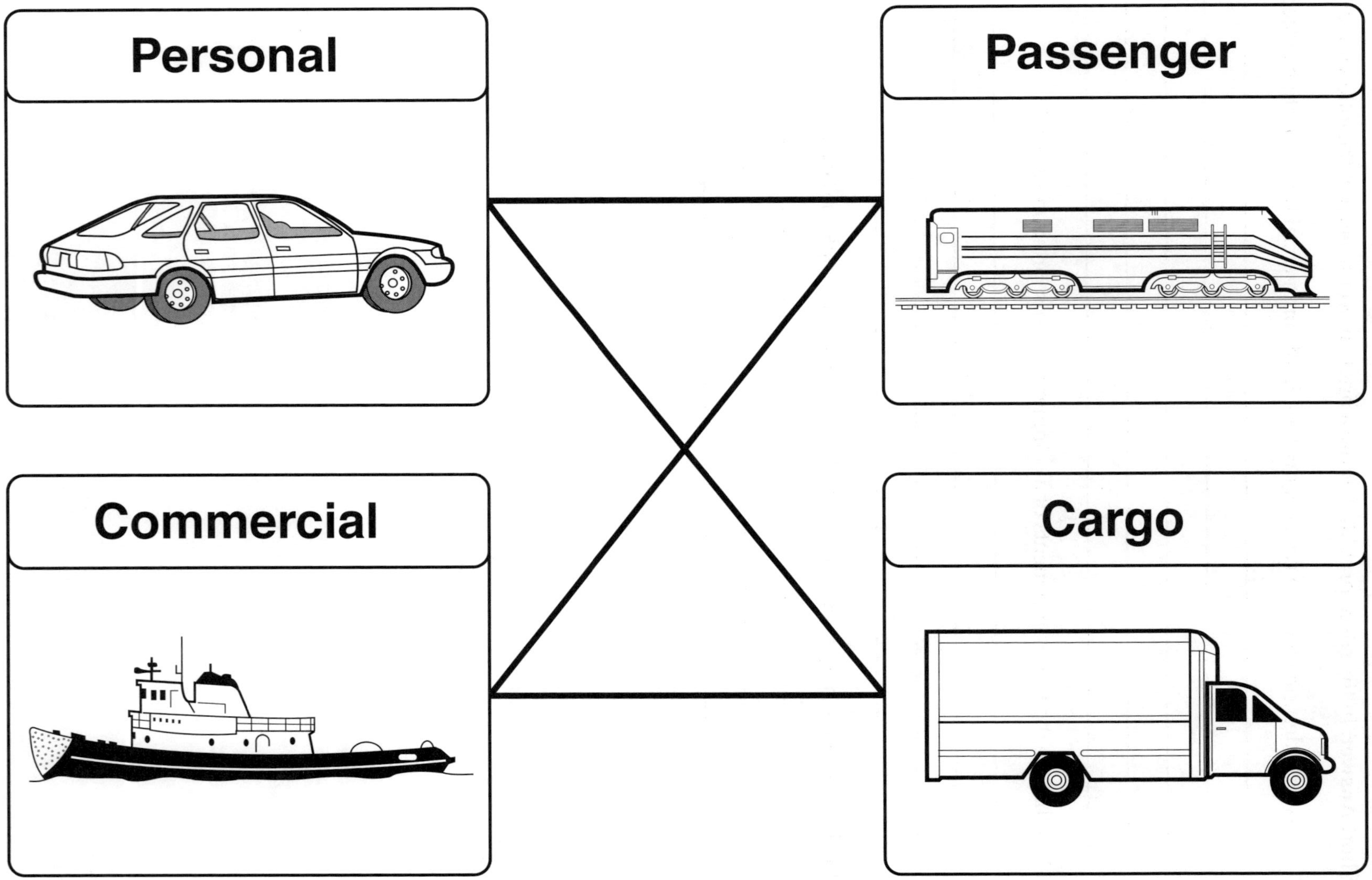

Transportation Processes

Routing: Determining the path the cargo or passengers will take.

Loading: Placing the cargo or allowing the passengers onto the vehicle.

Moving: Changing the geographic location of cargo or passengers.

Unloading: Removing the cargo from or allowing passengers to leave the vehicle.

Storing: Placing the cargo into a secure and/or protected location.

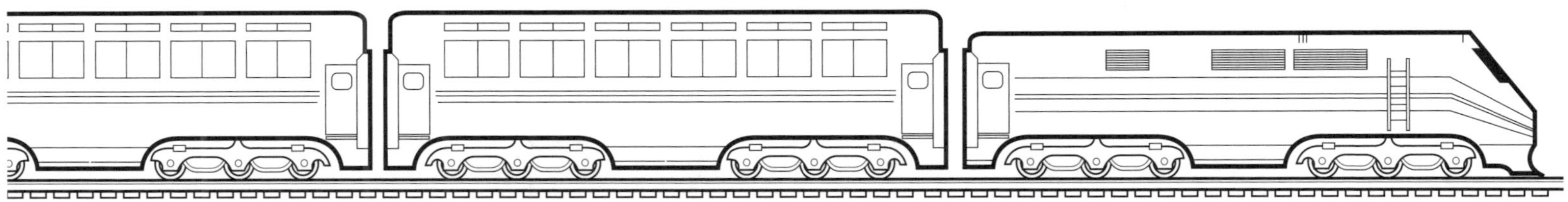

Section 8

Applying Technology: Using Energy

Teaching Materials

Text

Chapter 27, pp. 526–537
  Test Your Knowledge, pp. 535–536
  Applying Your Knowledge, p. 537

Chapter 28, pp. 538–561
  Test Your Knowledge, p. 561
  Applying Your Knowledge, p. 561

Section 8 Activities
  Activity 8A—Design Problem, p. 562
  Activity 8B—Fabrication Problem, pp. 562–563

Student Activity Manual

Activity 8A—Design Problem, pp. 95–98
Activity 8B—Fabrication Problem, pp. 99–100
Taking Your Technology Knowledge Home, pp. 101–103

Teacher's Resources

Chapter 27 Quiz, pp. 262–263
Chapter 28 Quiz, pp. 268–269
Reproducible Masters
  27-1: Types of Energy, p. 264
  27-2: Forms of Energy, p. 265
  28-1: Energy Converters, p. 270
  28-2: Energy Conversion, p. 271
  28-3: Creating Motion, p. 272

Transparencies (Binder/CD only)
  27-1: Energy—Key Terms
  27-2: Energy Resources

Organization of Section 8

Energy is fundamental to all technological activities. It involves converting one of six basic types of energy into heat, light, sound, or motion, Figure 8-1. This section is designed to introduce students to:

- The concepts of energy and work
- The types of energy and energy sources
- Common energy converters

As the students complete Section 8, they will first learn about the types of energy. They will be able to describe energy as either the force doing the work or a force that has the ability of doing work. This will introduce them to both potential energy (stored) and kinetic energy (created by motion). Then they will learn that work means applying a force to move a mass a certain distance.

Using this foundation, the students will explore the six major forms of energy:

- Mechanical: Energy produced by motion of technological devices and natural forces, such as wind and falling water
- Radiant: Energy given off by hot objects
- Chemical: Energy stored with a chemical compound or complex substance
- Thermal: Energy related to heat (rapid movement of atoms or molecules in a substance)
- Electrical: Energy associated with electrons moving along a conductor
- Nuclear: Energy associated with the internal bonds of atoms

Finally, the students will investigate four major types of energy conversion systems:

- Inexhaustible energy converters
- Renewable energy converters
- Thermal energy converters
- Electrical energy converters

Section 8 includes content contained in two chapters and one laboratory activity. The chapters are:

- Chapter 27—Energy: The Foundation of Technology
- Chapter 28—Energy Conversion Systems

As with the other learning sections in this program, two different types of activities are provided. If you want students to address the content from a problem-solving approach, select Activity 8A—Design Problem. A more structured approach is provided in Activity 8B—Fabrication Problem, which allows students to build and test a specific energy converter.

Presenting Section 8

Estimated length: 3 weeks

Section 8 is designed to help students understand that:

- Energy is fundamental to all technologies.
- Energy sources may be exhaustible, renewable, or inexhaustible.

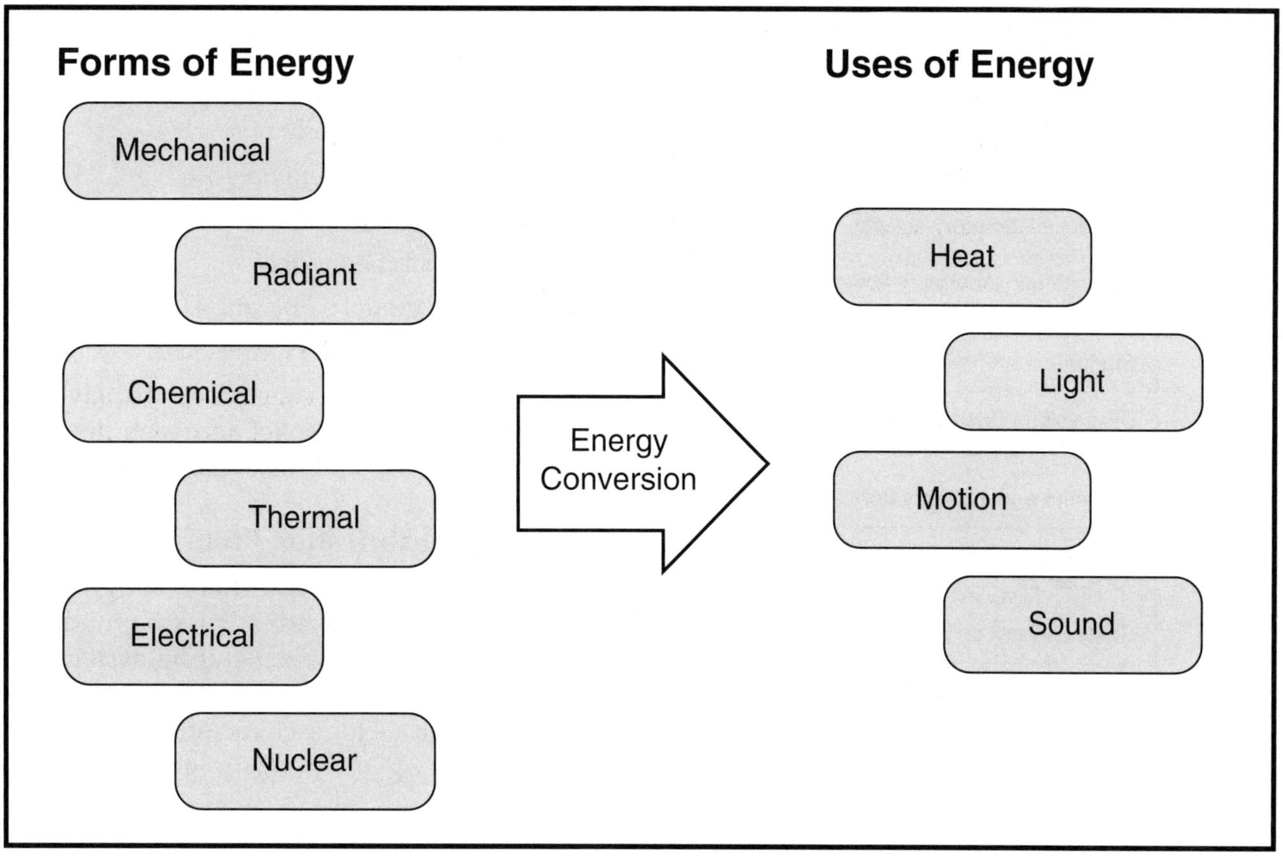

Figure 8-1

- Energy is constantly being converted in form as it is applied to agriculture, communication and information processing, construction, manufacturing, medicine, and transportation systems.
- Conservation of energy is important.

The teaching strategies for this section include lectures and discussions to present content and realistic laboratory activities to reinforce these presentations. The fifteen-day schedule, shown in Figure 8-2, has been developed for this section. To prepare for presenting this section, you should:

1. Read Chapters 27 and 28 in *Technology*.
2. Prepare discussions on:
   a. Energy: The foundation of technology (Chapter 27).
   b. Energy conversion systems (Chapter 28).
   c. Energy conservation (Chapter 28).
3. Select a laboratory activity.
   a. Activity 8A—This is an open-ended problem-solving activity in which students design, build, and evaluate an energy converter.
   b. Activity 8B—This is a structured activity in which students build and test a wind-powered electric generation system.
4. Practice the selected laboratory activity.
5. Prepare teaching materials (transparencies or bulletin boards, for example) to support the content.
6. Determine which study questions and quiz questions will be used.
7. Review the Taking Your Technology Knowledge Home activity in the *Student Manual*.

Section 8: Energy Conversion Technology	
Day	Activity
1	Discussion: "Energy: The Foundation of Technology" (Chapter 27) Introduce laboratory activity
2–4	Demonstrate laboratory activity procedures Students work on laboratory activity
5–8	Discussion: "Energy Conversion Systems" (Chapter 28) Students work on laboratory activity
9–13	Discussion: "Energy Conservation Systems" (Chapter 28) Students work on laboratory activity
14	Students present results of laboratory activity
15	Summarize section

Figure 8-2

Presenting the Activities

Consider the following recommendations in presenting the activity you select for this section.

Activity 8A—Design Problem

This activity is appropriate for either individual or group work. The students are to design, build a prototype, and evaluate a device that will convert rotating motion produced by a horizontal waterwheel into reciprocating motion for a colonial sawmill or pounding motion for a forge. To complete this task they will be given the following challenge:

Select one of the two models listed above (reciprocating motion for a colonial sawmill or pounding motion for a forge). Design and build a working model using wood, cardboard, or other easily worked materials.

During the activity each student or group of students will:

1. Generate rough sketches of three devices that will meet the design challenge.
2. Select the best idea represented on the sketches.
3. Refine the best idea and present the idea on a refined sketch.
4. Dimension the refined sketch.
5. Prepare a materials list for the model.
6. Construct the model.
7. Test the model's operation.
8. Present the model to the class.
9. If time permits, construct a display that focuses on the model and such items as its operation and historical importance.

Activity 8B—Fabrication Problem

During this activity, each student or group of students will build and test a working model of a wind-powered electric generation system. To complete the activity, the students will need the following materials for each model:

- One piece ¾" × 5½" × 8" wood base (pine, fir, spruce)
- One strip ¾" × 18" strip of thin gage sheet metal
- Two ½" × #6 sheet metal screws
- One small electric motor (available from a hobby or electronics store)
- One multimeter
- 24" electrical hook-up wire
- One 3-blade or 4-blade propeller (available from a hobby store)
- During the activity, each student should:
  1. Produce a dimensioned sketch of the model she or he will build.
  2. Develop a materials list for the important parts of the model.
  3. Observe a demonstration of the techniques that will be used to make the model. **Note:** During the demonstration, each student should list each operation, the machine used for it, and any safety precautions he or she observed while completing the operation.
  4. Follow the procedures listed in *Technology* to:

a. Build the prototype.
b. Test the prototype.
   Note: During the testing, the student may want to:
   a. Use a fan with a different number of blades to repeat the experiment and then observe and explain the differences in readings from the original experiment.
   b. Change the angle the wind is hitting the propeller to 30° off center and repeat the experiment. Observe and explain the differences in readings from the original experiment.

Reinforcing and Enriching

The following activities can be used to supplement and enrich the laboratory activities at the end of each textbook section and the Applying Your Knowledge activities included at the end of each textbook chapter.

1. If time permits, the section can be enhanced by having the students complete both activities. Another approach is to have half of the class complete the design problem and have the other half complete the fabrication problem. At the end of the section, each group could share its experiences with the other group.
2. The design problem could be based on an energy conversion problem developed by the students. This approach would build on students' interests.

Taking Your Technology Knowledge Home

This take-home activity allows students to apply their knowledge of energy and energy conservation to daily life. They will:

1. Survey several rooms in their home and identify the type of lighting used.
2. Record the wattage of each bulb used.
3. Estimate the number of hours each bulb is used in a month.
4. Calculate the cost of operating each bulb for one month.
5. Identify bulbs that can be replaced with compact fluorescent bulbs.
6. Recalculate the cost of operating the light for the month.
7. Determine the possible monthly savings that could be achieved using compact fluorescent bulbs.
8. Determine the cost of exchanging the bulbs.
9. Determine the number of months it would take for the savings to pay for the bulbs.

Adjustments for Exceptional Students

Either activity in this section can be completed in relatively short time. Adjustments for students with special needs can be made in several ways. One approach would be to adjust the level of sophistication for the design and fabrication problems. Gifted students could be challenged to provide more creative answers and possibly produce a more complex solution. Lower achievers can still complete either activity but with solutions reflecting their ability levels.

Also, the level of structure provided for the activity can be adjusted. Often, lower-ability students need more structure and shorter, more frequent demonstrations and presentations. They may require reinforcement almost daily. Guard against providing too much structure, however, because this often stifles creativity and is less appealing to gifted students.

Finally, having groups complete the activities allows students of all abilities to contribute their unique talents. This approach requires careful grouping so that individual abilities complement one another. Some students will possess more abstract reasoning ability, others will have more artistic abilities, and still others might have more manual dexterity. Group work allows students to develop cooperative work habits and to contribute their own unique abilities to the team.

Evaluation Procedures

The understanding that students have of technology and technological systems should be measured in several ways, including:

- Observation and critical listening during class presentations and discussions.
- Worksheets and quizzes covering the cognitive material presented.
- Open-ended questions about the learned material and how technology is developed.
- Assessment of individual participation and contributions to group work.
- Assessment of the quality and creativeness of the solutions to the assigned design/ fabrication activities.

Chapter 27
Energy: The Foundation of Technology

Objectives

After studying Chapter 27, the students will be able to:

- Define energy.
- Describe the difference between potential and kinetic energy.
- Identify the differences among energy, work, and power.
- Identify the six major forms of energy.
- Identify exhaustible, renewable, and inexhaustible energy sources.
- Explain how energy technology can make our lives better and how energy technology can cause damage.

Answers to Test Your Knowledge *Questions, Text Pages 535–536*

1. Energy is the ability to do work.
2. Kinetic energy is a force that is doing work. Potential energy is a force that has the potential to do work.
3. b. Work
4. c. Power
5. b. Work
6. a. Energy
7. c. Power
8. a. Energy
9. b. Work
10. True
11. The sun
12. False
13. Fusion
14. True
15. b. Renewable
16. c. Inexhaustible
17. b. Renewable
18. c. Inexhaustible
19. a. Exhaustible
20. a. Exhaustible
21. c. Inexhaustible
22. b. Renewable
23. a. Exhaustible
24. Evaluate individually.
25. Evaluate individually.

Answers to the Chapter Quiz, Teacher's Resource

1. True
2. False
3. True
4. True
5. False
6. False
7. False
8. False
9. True
10. c. Drawbar horsepower
11. a. Thermal energy
12. d. Thermal energy
13. Energy is defined as the ability to do work.
14. Work is applying a force that moves a mass a distance in the direction of the applied force.
15. Evaluate individually.

Captions for Reproducible Masters

Reproducible Master 27-1. Types of Energy—Shown are the two types of energy. Stored energy is potential energy. Energy in motion is kinetic energy.

Reproducible Master 27-2. Forms of Energy—Listed are the six forms into which all types of energy can be grouped.

Captions for Transparency Masters (Binder/CD Only)

Transparency Master 27-1. Energy/Key Terms—Work and power are important concepts when discussing energy. Work applies a force to move a mass over a distance. Power measures the rate at which work is done.

Transparency Master 27-2. Energy Resources—Shown are the three types of energy resources with examples of each.

Energy: The Foundation of Technology

Name ______________________________ Period ____________

Date ______________________________ Score ____________

Chapter 27 Quiz

True/False: If a statement is true, write *True* in the blank. If a statement is false, write *False* in the blank.

_________ 1. Unburned coal is an example of potential energy.

_________ 2. Power is defined as applying a force to move a mass a certain distance.

_________ 3. Corn and coal are both sources of chemical energy.

_________ 4. Hydroelectric generators, wind-powered generators, and solar panels harness the energy of the solar weather system.

_________ 5. Low-frequency waves contain more energy than high-frequency waves of the same amplitude.

_________ 6. A horsepower is the force that is needed to move 500 pounds one foot in one minute.

_________ 7. Combining two atoms into a new larger atom is called fission.

_________ 8. The horsepower a machine can generate under ideal conditions is called the drawbar horsepower.

_________ 9. The two types of nuclear reactions are fission and fusion.

Multiple Choice: Choose the best answer and write the corresponding letter in the blank.

_____ 10. Power deliver to the hitch of tractors is called:
- a. brake horsepower.
- b. frictional horsepower.
- c. drawbar horsepower.
- d. indicated horsepower.

_____ 11. Heat energy is also referred to as:
- a. thermal energy.
- b. radiant energy.
- c. chemical energy.
- d. nuclear energy.

_____ 12. The internal movement of atoms in a substance creates:
- a. nuclear energy.
- b. radiant energy.
- c. chemical energy.
- d. thermal energy.

(Continued)

Name ____

Short Answer: In the space provided, write a brief answer to each of the following questions.

13. Define *energy.*

____

____

____

____

____

14. Define *work.*

____

____

____

____

____

15. Compare the benefits and drawbacks of one type of energy technology.

____

____

____

____

____

Types of Energy

Potential

Stored Energy

Kinetic

Energy in Motion

Forms of Energy

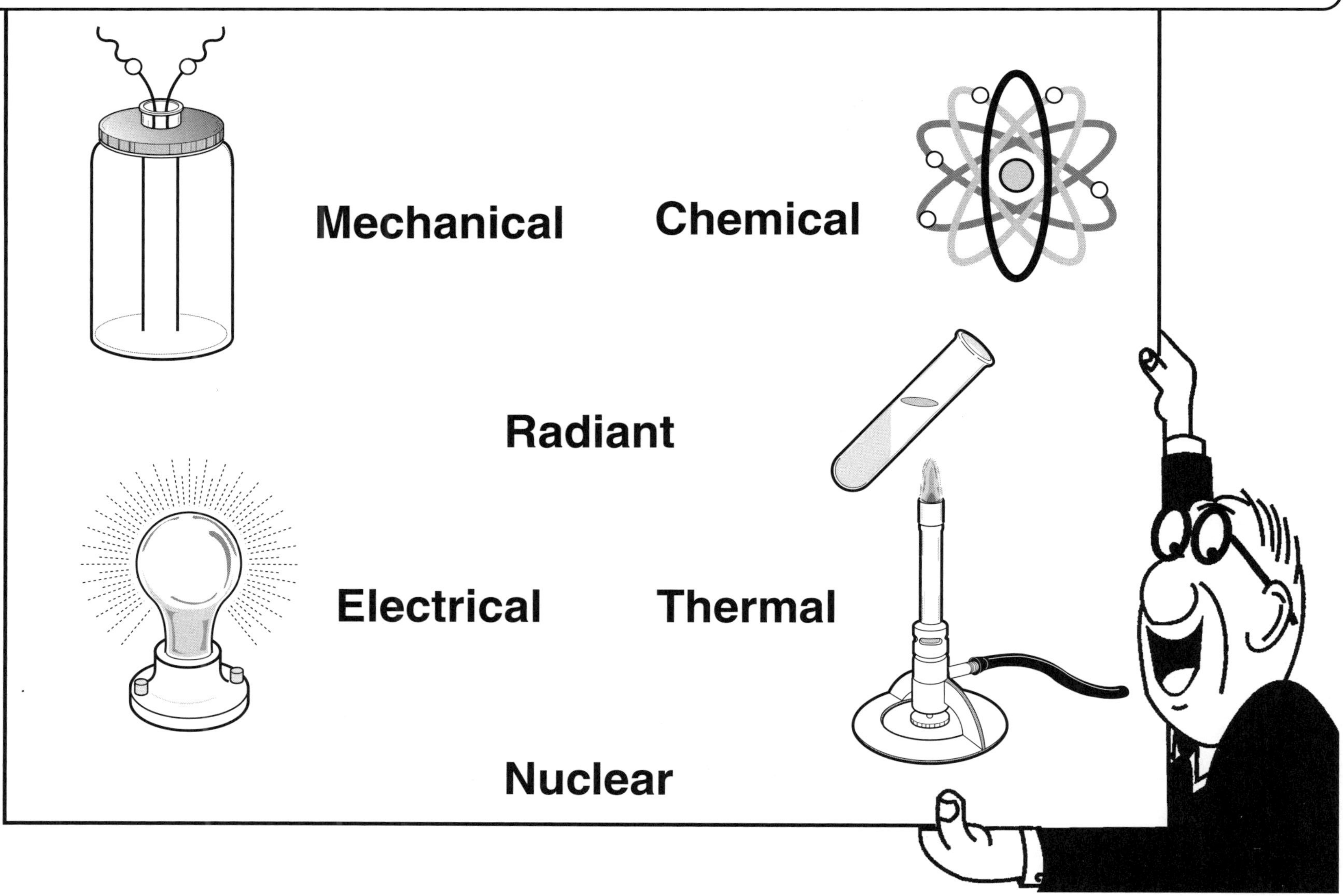

Chapter 28
Energy Conversion Systems

Objectives

After studying Chapter 28, the students will be able to:

- Describe devices that convert inexhaustible energy into mechanical motion.
- Describe the main ways solar energy is converted into other forms of energy.
- Explain the differences between passive and active solar conversion systems.
- Describe a common geothermal energy conversion system.
- Explain the operation of a common biomass converter.
- Describe heat engines in terms of energy conversion.
- Explain the differences between internal and external combustion engines.
- Describe common ways to heat homes and buildings.
- Describe the major parts of an electric energy generation and conversion system.
- Describe the common energy input systems for electric generation plants.
- Explain how energy is applied to work.

Answers to Test Your Knowledge *Questions, Text Page 561*

1. True
2. Evaluate individually. Common answers include windmills and sailboats.
3. Passive collectors directly collect, store, and distribute the heat they convert from solar energy. Active collectors use pumps to circulate the water that collects, stores, and distributes the heat they convert from solar energy.
4. Evaluate individually.
5. Student answers may include either ocean thermal energy conversion systems or ocean mechanical energy conversion systems.
6. Biochemical
7. Evaluate individually.
8. Student answers may include conduction, convection, or radiation.
9. True
10. Fluid power

Answers to the Chapter Quiz, Teacher's Resource

1. True
2. True
3. False
4. False
5. True
6. True
7. a. The sun
8. c. Ocean thermal conversion
9. b. Direct combustion
10. c. The transfer of heat between or within fluids
11. d. Transformers
12. b. Levers
13. a. Geothermal heat pumps
14. Evaluate individually. Common answers include windmills, sailboats, and waterwheels.
15. Passive collectors directly collect, store, and distribute the heat they convert from solar energy. Active collectors use pumps to circulate the water that collects, stores, and distributes the heat they convert from solar energy.

Captions for Reproducible Masters

Reproducible Master 28-1. Energy Converters—Energy of all forms can be converted into heat, light, sound, and motion.

Reproducible Master 28-2. Energy Conversion—Shown is the process that turns fuel into electricity in a steam power plant.

Reproducible Master 28-3. Creating Motion—Shown are three important types of motion along with mechanisms that can be used to produce these motions.

Energy Conversion Systems

Name ______________________________ Period ______________

Date ______________________________ Score ______________

Chapter 28 Quiz

True/False: If a statement is true, write *True* in the blank. If a statement is false, write *False* in the blank.

_________ 1. Humans cannot create energy but can convert it to meet their needs.

_________ 2. Today's wind turbine is used primarily to pump water for livestock.

_________ 3. Passive collectors use pumps to circulate the water that collects, stores, and distributes the heat they convert from solar energy.

_________ 4. Fermentation is a controlled decaying process that takes place without oxygen.

_________ 5. Radiant heating uses high-resistance wires in the ceiling that radiate warmth into a room.

_________ 6. Cams are used to convert rotating motion into reciprocating motion.

Multiple Choice: Choose the best answer and write the corresponding letter in the blank.

_____ 7. The source of most energy on earth is:
   a. the sun.
   b. the wind.
   c. water.
   d. None of the above.

_____ 8. Which of the following types of ocean energy technologies can be used only at the equator?
   a. Wave energy
   b. Tidal power
   c. Ocean thermal conversion
   d. Both b. and c.

_____ 9. Which of the following thermochemical conversion methods is most commonly used for biofuels?
   a. Pyrolysis
   b. Direct combustion
   c. Liquefaction
   d. None of the above

_____ 10. Convection is defined as:
   a. the movement of heat along a solid material.
   b. heat transfer using electromagnetic waves.
   c. the transfer of heat between or within fluids.
   d. Both a. and c.

(Continued)

Name ______________________________

_____ 11. Voltage of electricity is changed by:
a. turbines.
b. commutators.
c. field magnets.
d. transformers.

_____ 12. Which of the following devices change direction of a linear force?
a. Cranks
b. Levers
c. Gears
d. Both a. and b.

_____ 13. Which of the following uses the constant 55°F temperature of ground water to heat homes?
a. Geothermal heat pumps
b. Passive solar collectors
c. Photovoltaic cell
d. None of the above

Short Answer: In the space provided, write a brief answer to each of the following questions.

14. List two early energy converters that used inexhaustible energy resources.

____

____

15. Describe the difference between active and passive solar energy conversion systems.

____

____

____

____

____

Energy Converters

Energy cannot be created or destroyed. It can be converted from one form to another to meet human needs.

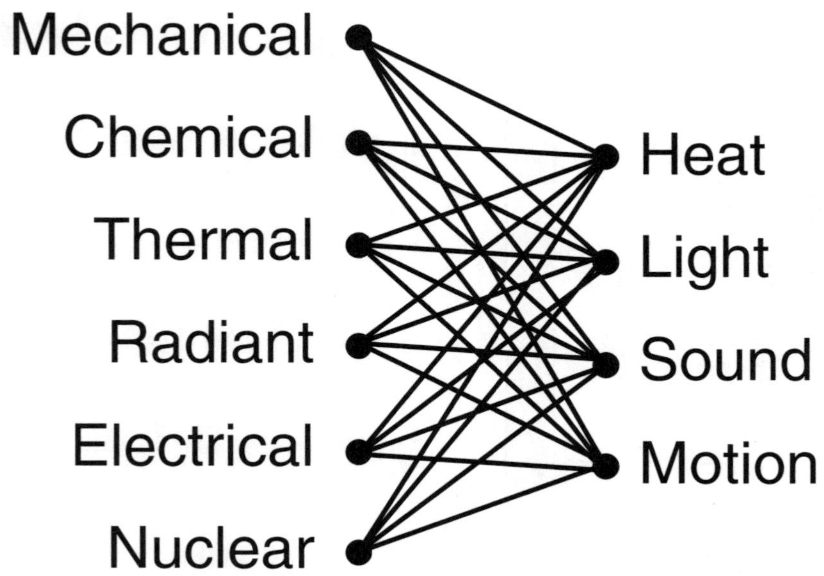

Energy Conversion
Steam Power Plant

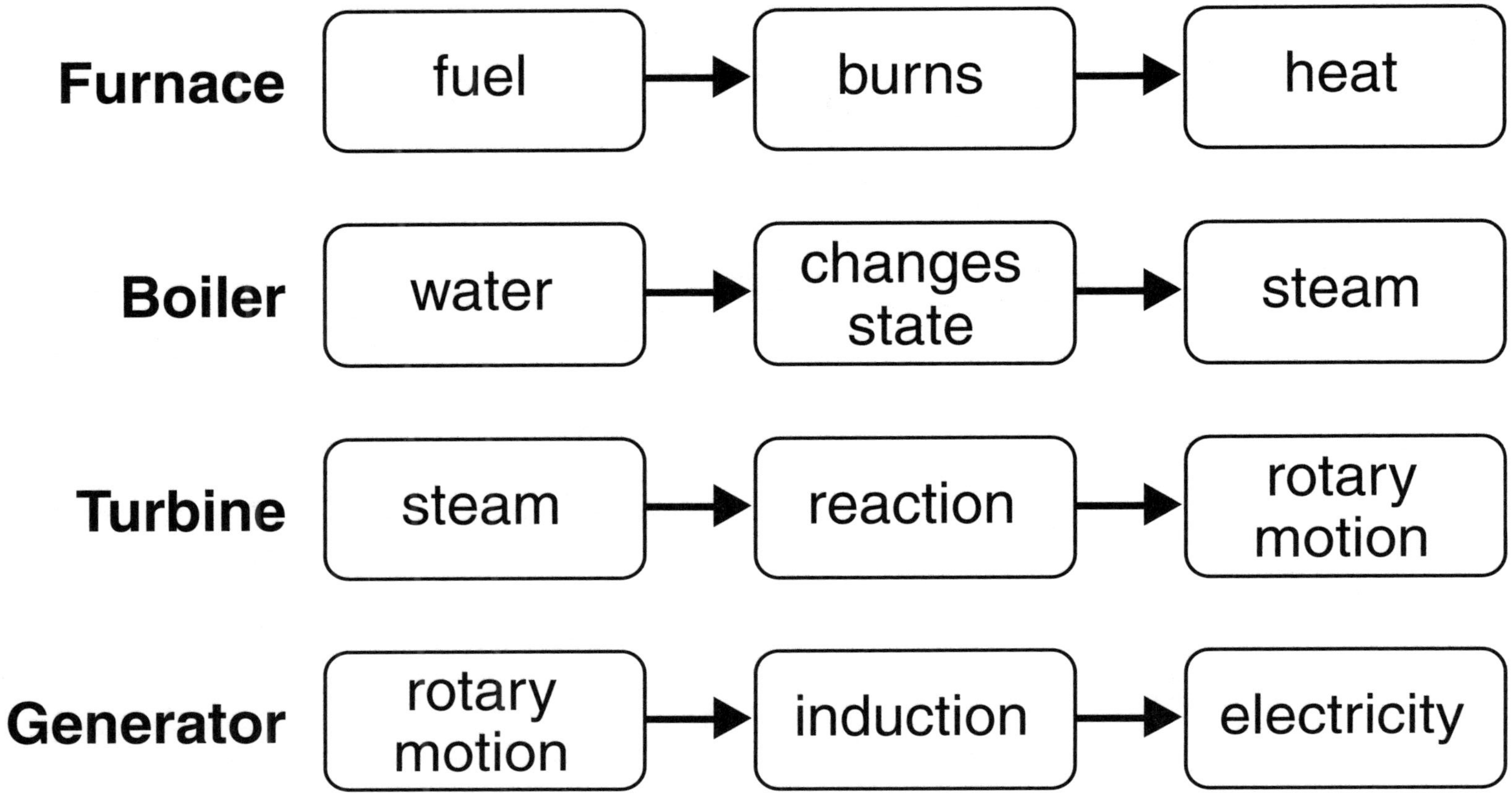

Creating Motion

Types	Mechanisms
Linear	Wheels
Reciprocating	Cams
Rotary	Screws
	Cranks
	Levers and Linkages

Section 9

Applying Technology: Meeting Needs through Biorelated Technologies

Teaching Materials

Text

Chapter 29, pp. 566–591
  Test Your Knowledge, pp. 588–589
  Applying Your Knowledge, p. 589

Chapter 30, pp. 592–613
  Test Your Knowledge, pp. 611–612
  Applying Your Knowledge, p. 613

Chapter 31, pp. 614–635
  Test Your Knowledge, pp. 632–633
  Applying Your Knowledge, p. 633

Section 9 Activities
  Activity 9A—Design Problem, p. 636
  Activity 9B—Fabrication Problem, pp. 636–637

Student Activity Manual

Activity 9A—Design Problem, pp. 107–109
Activity 9B—Fabrication Problem, p. 111
Taking Your Technology Knowledge Home, pp. 113–115

Teacher's Resources

Chapter 29 Quiz, pp. 281–282
Chapter 30 Quiz, pp. 288–289
Chapter 31 Quiz, pp. 295–296
Reproducible Masters
  29-1: Major Farm Crops, p. 283
  29-2: Types of Farm Equipment, p. 284
  29-3: Parts of a Tractor, p. 285
  30-1: Types of Food Processing, p. 290
  30-2: Methods of Preserving Food, p. 291
  31-1: Factors in Wellness, p. 297
  31-2: Technology in Sports, p. 298
  31-3: Goals of Medicine, p. 299

Transparencies (Binder/CD only)
  29-1: Types of Agriculture
  30-1: Types of Primary Food Processes
  31-1: Focus of Health and Medicine

Organization of Section 9

Section 9 is designed to allow students to explore three major applications of technology as it applies to living things. These, as shown in Figure 9-1, are:

- Growing and harvesting food
- Processing food

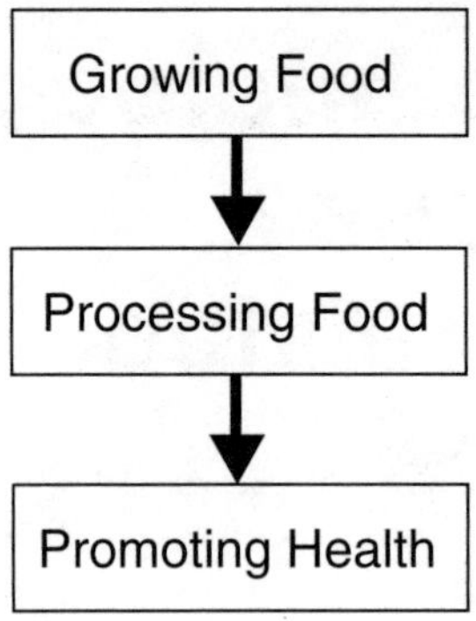

Figure 9-1

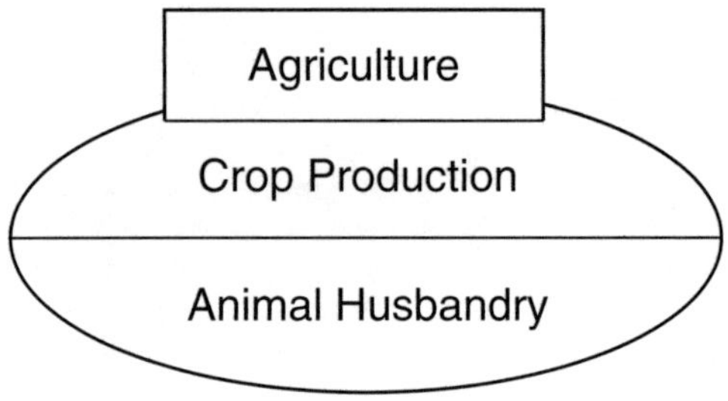

Figure 9-2

- Promoting human health and treating illnesses

The students will first investigate the field of agriculture. They will explore the two types of agriculture, Figure 9-2. These are crop production, which grows crops, and animal husbandry, which involves breeding, raising, and training animals. They will explore the types of crops that are grown on a variety of farms. These include grains, fruits, vegetables, nuts, and forage crops. They will also explore the major types of equipment that have been developed to aid in farming. These include pulling, tillage, planting, pest control, irrigation, harvesting, transporting, and storing equipment. The study of agriculture will conclude with discussions of the technology used in raising livestock and the relationship between agriculture and biotechnology.

The second area of study in Section 9 is food processing. It will allow students to investigate the technology used in preparing and processing foods. It will focus on the two types of food processing shown in Figure 9-3:

- Primary processing
- Secondary processing

Raw agricultural materials are converted by

Primary Processing

into food commodities or ingredients that can be converted by

Secondary Processing

into edible products.

Figure 9-3

They will discover that food is processed by material conversion and food preservation techniques. They will learn that agricultural products are processed into food using mechanical, thermal, and chemical processes. They will read about flour milling, coffee roasting, and cheese making as examples of these techniques. Later they will explore drying, curing, canning, aseptic packaging, refrigerating and freezing, controlled atmospheric storing, fermenting, pasteurizing, irradiating, and adding preservatives as major ways of preserving food products. Finally, they will learn about secondary food processing and how a new food product is developed.

The last area of study in Section 9 is medical and health technologies. In this area the students will learn about two areas of this topic:

- Wellness: a state of physical well-being
- Illness: a state of poor health

Students will then explore how exercise and sports have been developed to promote wellness. Next, they will explore the types of technology that have been developed to diagnose and treat illnesses and medical conditions. Finally, they will explore how new drugs and vaccines are developed and tested.

For this section, the students spend ten days covering the content of three chapters and in completing two laboratory activities. These chapters are:

- Chapter 29—Agricultural and Related Biotechnologies
- Chapter 30—Food-Processing Technologies
- Chapter 31—Medical and Health Technologies

This section, like all others in this course, has two different types of activities provided. The design problem allows students to engage in a creative, problem-solving approach, and the fabrication problem provides a more structured approach to the laboratory experience.

Presenting Section 9

Estimated length—2 weeks

Section 9 helps students understand several areas where technology is applied to living things. Specifically this section deals with:

- Agriculture
- Food processing
- Medicine and health care

The content of Section 9 could be presented solely using lecture/discussion strategies. Technology education, however, is based on the concept that presentations should be reinforced with realistic laboratory activities. Thus, about 75 percent of the time assigned to this section is dedicated to laboratory activity. The ten-day schedule developed for this section reflects this emphasis, Figure 9-4.

Section 9: Applying Technology: Meeting Needs through Biorelated Technologies	
Day	Activity
1–3	Discussion: "Agricultural and Related Technologies" (Chapter 29) Complete laboratory activity 9B
4–7	Discussion: " Food-Processing Technologies" (Chapter 30) Introduce laboratory activity 9A
8–9	Discussion: "Medical and Health Technologies" (Chapter 31) Complete laboratory activity 9A
10	Summarize section

Figure 9-4

Three important topics are introduced through presentations and discussions. These are (1) agricultural and related biotechnologies, (2) food-processing technologies, and (3) medical and health technologies. The students will be using simple tools to complete both a design problem and a fabrication problem. It is important that the students use tools and machines SAFELY. Demonstrate appropriate safety procedures prior to any laboratory activity.

To prepare to present this section, you should:

1. Read Chapters 29, 30, and 31 in *Technology.*
2. Prepare discussions on:
   a. Agricultural and related biotechnologies (Chapter 29).
   b. Food-processing technologies (Chapter 30).
   c. Medical and health technologies (Chapter 31).
3. Review the laboratory activities.
   a. Activity 9A—This activity is a creative, problem-solving activity.
   b. Activity 9B—This activity is a structured activity. Groups of students will use a food dehydrator to preserve an agricultural product.
4. Practice the selected laboratory activity.
5. Prepare teaching materials (transparencies or bulletin boards, for example) to support the content.
6. Determine which study questions and quiz questions will be used.

Presenting the Activities

Consider the following recommendations in presenting the activity you select for this section.

Activity 9A—Design Problem

In this activity, the students are challenged to design a device to aid a person with a medical condition. They should be provided a kit of common materials, such as wood strips, dowels, rubber bands, thread spools, small nuts, paper fasteners, small wooden blocks, and string.

They are then asked to develop the technological device using these materials.

To present this activity:

1. Introduce the background, situation, and challenge of the activity.
2. Divide the class into groups of two to four students. **Note:** Students may work independently on this activity.
3. Have each member of each group:
   a. Work independently to develop at least two solutions to the problem.
   b. Present his/her solution to their group.
4. Have each group:
   a. Discuss each solution.
   b. Select a "best" solution.
   c. Develop a supply list for the product.
   d. Construct at least one working model of the technological device using the supplies listed.
   e. Present their solution to the class.

Activity 9B—Fabrication Problem

To present this activity:

1. Discuss the background and challenge for this activity, which is included in the Section Activity in the textbook. The challenge for the students is to use a food dehydrator to preserve a food product for later use.
2. Divide the class into groups of two or three students.
3. Have the groups follow the procedures listed on the activity sheet to complete the activity.

Reinforcing and Enriching

The following activities can be used to supplement and enrich the laboratory activities at the end of each textbook section and the Applying Your Knowledge activities included at the end of each textbook chapter.

1. Have the students develop a package for their dehydrated product. The package should include appropriate graphics, nutritional information, and other texts.
2. Prepare a user's manual for their biomechanical product. It should tell the

product user how to use, care for, and maintain their products.

Taking Your Technology Knowledge Home

This take-home activity allows students to apply their knowledge of food processing at home. They will:

1. Use primary food-processing techniques to make peanut butter.
2. Use secondary food-processing techniques to make biscuits.
3. Prepare a snack using the products of their efforts.

Adjustments for Exceptional Students

Both laboratory activities are designed to be done in a short period of time. The design problem is open-ended to accommodate varying abilities and levels of sophistication. The fabrication problem is fairly simple, and its directions can be followed easily. By using carefully structured groups for each activity, individual differences in student abilities can be accommodated.

Evaluation Procedures

The understanding that students have of technology and technological systems should be measured in several ways, including:

- Observation and critical listening during class presentations and discussions.
- Worksheets and quizzes covering the cognitive material presented.
- Open-ended questions about the learned material and how technology is developed.
- Assessment of individual participation and contributions to group work.
- Assessment of the quality and creativeness of the solutions to the assigned design/fabrication activities.

Chapter 29
Agricultural and Related Biotechnologies

Objectives

After studying Chapter 29, the students will be able to:

- Define agriculture.
- Explain the difference between science and technology in agriculture.
- List the major types of agriculture.
- Name major crops grown on farms.
- Identify major equipment used in crop production.
- Describe how technology is used in hydroponics.
- List the types of livestock farms.
- Describe the technology used in raising livestock.
- Describe how technology is used in aquaculture.
- Define biotechnology.
- Describe ways that biotechnology can be used in agriculture.

Answers to Test Your Knowledge *Questions, Text Pages 588–589*

1. Agriculture is people using materials, information, and machines to produce the food and natural fibers.
2. Evaluate individually.
3. Crop production and animal husbandry
4. Grains
5. Forage
6. Storing
7. False
8. Plow
9. True
10. True
11. Sprinkler irrigation
12. Evaluate individually.
13. Hydroponics
14. Student answers may include: cattle ranches, dairies, and swine, horse, and poultry farms. Additional answers may be accepted at the instructor's discretion.
15. Student answers may include: rail, barbed wire, woven wire, and electric.
16. Silos are used to contain and protect grain and hay (silage).
17. Slurry
18. Aquaculture
19. Biotechnology is using biological agents in processes to produce goods or services.
20. Evaluate individually.

Answers to the Chapter Quiz, Teacher's Resource

1. False
2. False
3. True
4. False
5. True
6. False
7. True
8. True
9. False
10. True
11. b. They sink less and have less slippage.
12. c. Moldboard plows
13. a. A grain drill is the most widely used type of seed planter.
14. b. A solid-set sprinkler system
15. Biotechnology can be defined as using biological agents in processes to produce goods or services.

Captions for Reproducible Masters

Reproducible Master 29-1. Major Farm Crops—Shows forage crops, grains, fruits and berries, nuts, vegetables and nonfood crops as the major types of farm crops.

Reproducible Master 29-2. Types of Farm Equipment—Lists the types of equipment as pulling, tillage, planting, pest control, irrigation, harvesting, transportation, and storage equipment.

Reproducible Master 29-3. Parts of a Tractor—Lists the six parts of a tractor as a power source, transmission and drive train, wheels or tracks, speed and direction control, operator's areas, and hitch.

Caption for Transparency Master (Binder/CD Only)

Transparency Master 29-1. Types of Agriculture—Shows and defines crop production and animal husbandry.

Agricultural and Related Biotechnologies

Name ______________________________ Period ______________

Date ______________________________ Score ______________

Chapter 29 Quiz

True/False: If a statement is true, write *True* in the blank. If a statement is false, write *False* in the blank.

_________ 1. Edible grains are the end product of all types of crop production agriculture.

_________ 2. A modern farmer needs to know a great deal about technology but very little about science.

_________ 3. The United States uses more than 40 percent of its total land for farming activities.

_________ 4. Vegetables are the most widely grown crops in North America.

_________ 5. Wastewater is waste that contains less than 2 percent solids.

_________ 6. Aquaculture is also called capture fishing.

_________ 7. Special harvesting machines have been created for many different crops.

_________ 8. More than 570 patents have been issued for different designs of barbed wire.

_________ 9. Catalysts are altered or consumed by the reactions they cause.

_________ 10. During World War II, scientists used the fermentation process to produce antibiotics.

Multiple Choice: Choose the best answer and write the corresponding letter in the blank.

_____ 11. Which of the following is true of track-type tractors?
a. They are faster than wheel-type tractors.
b. They sink less and have less slippage.
c. They are not suitable for use in muddy conditions.
d. None of the above.

_____ 12. Which of the following types of plows uses a frame with several plowshares to cut and roll the soil?
a. Disc plows
b. Chisel plows
c. Moldboard plows
d. Both a. and c.

_____ 13. Which of the following best describes a grain drill?
a. A grain drill is the most widely used type of seed planter.
b. A grain drill is the most widely used type of grain elevator.
c. A grain drill is the most widely used type of harvester.
d. None of the above.

_____ 14. An irrigation system in which sprinkler lines are set up so the entire field can be irrigated all at once is known as:
a. a pivot sprinkler system.
b. a solid-set sprinkler system.
c. a unit-dispersion sprinkler system.
d. None of the above.

(Continued)

Name ____

Short Answer: In the space provided, write a brief answer to the following question.

15. Define *biotechnology.*

____

____

____

____

____

Major Farm Crops

Forage Crops

Grains

Fruits and Berries

Non-Food Crops

Nuts

Vegetables

Types of Farm Equipment

Pulling (tractors)

Tillage

Planting

Pest Control

Irrigation

Harvesting

Transportation

Storage

Parts of a Tractor

A Power Source

Transmission and Drive Train

Wheels or Tracks

Speed and Direction Control

Operator's Area

Hitch

Chapter 30
Food-Processing Technologies

Objectives

After studying Chapter 30, the students will be able to:

- Explain the role of food in daily life.
- Define food-processing technology.
- Describe the difference between primary and secondary food processing.
- Describe ways used to process foods.
- Identify ways to preserve foods.
- Describe the steps that might be used to develop a new food product.
- Describe the activities involved in manufacturing a new food product.

Answers to Test Your Knowledge *Questions, Text Pages 611–612*

1. Evaluate individually.
2. Food-processing technology involves using knowledge, machines, and techniques to convert agricultural products into foods that have specific textures, appearances, and nutritional properties.
3. Primary
4. Secondary
5. Grain receiving and cleaning, conditioning, milling, and storage and packaging
6. Drum or barrel
7. Student answers may include any of the following: receiving the milk, cooking, cutting the curd, making cheddar, salting, sealing and aging, and packaging.
8. i. Irradiation
9. c. Canning
10. a. Drying
11. f. Controlled atmosphere storage
12. j. Preservatives
13. b. Curing and smoking
14. e. Refrigeration and freezing
15. h. Pasteurization
16. d. Aseptic packaging
17. g. Fermentation
18. c. Canning
19. a. Drying
20. d. Aseptic packaging
21. True
22. Product modification
23. Pilot-plant production
24. Product specification
25. Storing, cleaning, sorting and grading, measuring, size reduction, compounding, and heat processing

Answers to the Chapter Quiz, Teacher's Resource

1. True
2. False
3. False
4. True
5. True
6. False
7. True
8. False
9. True
10. True
11. d. None of the above
12. c. Sensory evaluation
13. a. Blanching
14. b. Fermentation
15. Storing, cleaning, sorting and grading, measuring, size reduction, compounding, and heat processing

Captions for Reproducible Masters

Reproducible Master 30-1. Types of Food Processing—Lists and describes primary and secondary food processing.

Reproducible Master 30-2. Methods of Preserving Food—Lists canning, drying, curing and smoking, aseptic packaging, refrigerating and freezing, fermentation, controlled atmosphere storage, preservatives, irradiation, and pasteurization as ways to preserve food.

Caption for Transparency Master (Binder/CD Only)

Transparency Master 30-1. Types of Primary Food Processes—Shows and gives examples of mechanical, thermal, and chemical food processes.

Food-Processing Technologies

Name ____ Period ____________________

Date ____ Score ____________________

Chapter 30 Quiz

True/False: If a statement is true, write *True* in the blank. If a statement is false, write *False* in the blank.

_________ 1. Food sustains life by providing seven basic components.

_________ 2. Fats are substances people eat that their bodies cannot digest.

_________ 3. Primary food-processing technology is used to make finished food products.

_________ 4. Wheat flour is the most commonly produced flour in the Western world.

_________ 5. Coffee beans are found inside a red cherry-shaped fruit.

_________ 6. Air roasting is the most common way to roast coffee beans.

_________ 7. The canning process preserves food by heating it in a vacuum-sealed container.

_________ 8. Aseptic packaging alters a food's flavor, texture, and nutrient content more than canning.

_________ 9. Controlled atmospheres can extend the storage life of fruits and vegetables by several months.

_________ 10. The size reduction process involves trimming, slicing, and crushing the basic ingredients.

Multiple Choice: Choose the best answer and write the corresponding letter in the blank.

_____ 11. Assessing the quality of the ingredients to be used in making the food product is called:
a. compounding.
b. storing.
c. measuring.
d. None of the above.

_____ 12. The company asks trained personnel to comment on the appearance, taste, smell, and texture of the products during which phase of product development?
a. Pilot-plant production
b. Product modification
c. Sensory evaluation
d. Both a. and c.

_____ 13. Heating food with steam or water for a short period of time before canning is known as:
a. blanching.
b. pasteurizing.
c. irradiating.
d. None of the above.

(Continued)

Name ______________________________

_____ 14. Which process uses microorganisms to break down complex organic compounds into simpler substances?
a. Pasteurization
b. Fermentation
c. Curing
d. None of the above

Short Answer: In the space provided, write a brief answer to the following question.

15. List the seven steps involved in manufacturing a new food product.

____

____

____

____

____

____

____

Types of Food Processing

Primary Processing

Converting Agricultural Materials into Food Commodities

Secondary Processing

Converting Food Commodities into Edible Products

Methods of Preserving Food

Canning

Drying

Curing and Smoking

Refrigerating and Freezing

Aseptic Packaging

Fermentation

Controlled Atmosphere Storage

Irradiation

Pasteurization

Preservatives

Chapter 31
Medical and Health Technologies

Objectives

After studying Chapter 31, the students will be able to:

- Describe the roles of science and technology in wellness and illness.
- Name the four major factors involved in wellness.
- Explain how technology is applied to exercise and sports.
- Define medicine.
- Identify kinds of health care professionals.
- Describe how technological devices are used in diagnosing illnesses and physical conditions.
- Explain how diagnostic imaging systems work.
- Describe how technological devices are used to treat illnesses and physical conditions.
- Explain how a new drug is developed.
- Identify work done by biomedical engineers.
- Describe how technology is used in surgical procedures.

Answers to Test Your Knowledge *Questions, Text Pages 632–633*

1. Evaluate individually.
2. Nutrition and diet, environment, stress management, and physical fitness
3. Evaluate individually.
4. Evaluate individually.
5. Preventing
6. True
7. Student answers may include any of the following: scales, thermometers, devices to do such things as measure the oxygen in the blood, listen to heart rhythm (stethoscopes), and measure blood pressure.
8. A computerized tomography scan is three-dimensional, whereas a conventional X ray is two-dimensional.
9. The ultrasound machine subjects the body to high-frequency sound pulses using a probe. The sound waves travel into the body. There, they hit a boundary between tissues. This boundary could be between soft tissue and bone or between fluid and soft tissue. Some of the sound waves are reflected back to the probe. Others travel further until they reach another boundary and get reflected. The probe picks up the reflected sound waves. They are processed by a computer to produce a still (photograph) or moving (television) image.
10. An endoscope is a narrow flexible tube containing a number of fiber-optic fibers that allow a physician to look inside the body.
11. False
12. Proprietary drugs are those drugs considered safe for unsupervised use by consumers.
13. True
14. Student answers may include examples of: life support equipment, artificial body parts, computer systems to monitor patients, sensors for blood chemistry, instruments and devices for therapeutic uses, and medical imaging and treatment systems.
15. Student answers may include: ultrasound, lasers, or endoscopes. Other answers may be accepted at the instructor's discretion.

Answers to the Chapter Quiz, Teacher's Resource

1. True
2. False
3. False
4. False
5. False
6. True
7. False
8. True
9. True
10. True
11. a. Radiation therapy
12. d. All of the above
13. b. Proprietary drugs

14. b. Three
15. Evaluate individually.

Captions for Reproducible Masters

Reproducible Master 31-1. Factors in Wellness—Lists nutrition and health, environment, stress management, and physical fitness as factors in a wellness program.

Reproducible Master 31-2. Technology In Sports—Presents playing venues, game equipment, and personal protection and clothing as areas where technology is used in sports.

Reproducible Master 31-3. Goals of Medicine—Presents prevention, diagnosis, and treatment as the three goals of medicine.

Caption for Transparency Master (Binder/CD Only)

Transparency Master 31-1. Focus of Health and Medicine—Identifies wellness and illness as the foci of health and medicine.

Medical and Health Technologies

Name ______________________________ Period ______________

Date ______________________________ Score ______________

Chapter 31 Quiz

True/False: If a statement is true, write *True* in the blank. If a statement is false, write *False* in the blank.

_________ 1. Technology provides the tools and equipment needed to achieve wellness and treat illness.

_________ 2. Anaerobic exercise uses oxygen to keep large muscle groups moving continuously.

_________ 3. Aerobic exercise increases strength and muscle mass but provides limited benefit to cardiovascular health.

_________ 4. The term *disease* describes only illnesses caused by viruses.

_________ 5. Physicians are the only professional health care workers who can prescribe medication.

_________ 6. Treatment involves applying medical procedures to cure diseases, heal injuries, or ease symptoms.

_________ 7. X-ray images are useful only in diagnosing bone fractures.

_________ 8. X rays can damage the human body.

_________ 9. Many medical conditions, including cases of cancer, are diagnosed by removing a sample of tissue.

_________ 10. When a new drug is approved for clinical trials, the drug is first tested on healthy people.

Multiple Choice: Choose the best answer and write the corresponding letter in the blank.

_____ 11. Many types of cancer are treated using:
a. radiation therapy.
b. balloon angioplasty.
c. intervention radiology.
d. Both a. and b.

_____ 12. Which of the following is a type of prostheses?
a. Dentures
b. Artificial arms
c. Artificial hearts
d. All of the above

_____ 13. Drugs that are considered safe for unsupervised use by consumers are called:
a. ethical drugs.
b. proprietary drugs.
c. vaccines.
d. None of the above.

(Continued)

Name ______________________________________

_____ 14. FDA-approved clinical tests are completed in how many phases?
a. Two
b. Three
c. Four
d. The number of phases depends on the drug being tested.

Short Answer: In the space provided, write a brief answer to the following question.

15. Give one example of a newer technology being used in surgery.

____

____

____

____

____

Factors in Wellness

- Nutrition and Health
- Environment
- Stress Management
- Physical Fitness

Technology in Sports

- Playing Venue
- Game Equipment
- Personal Protection and Clothing

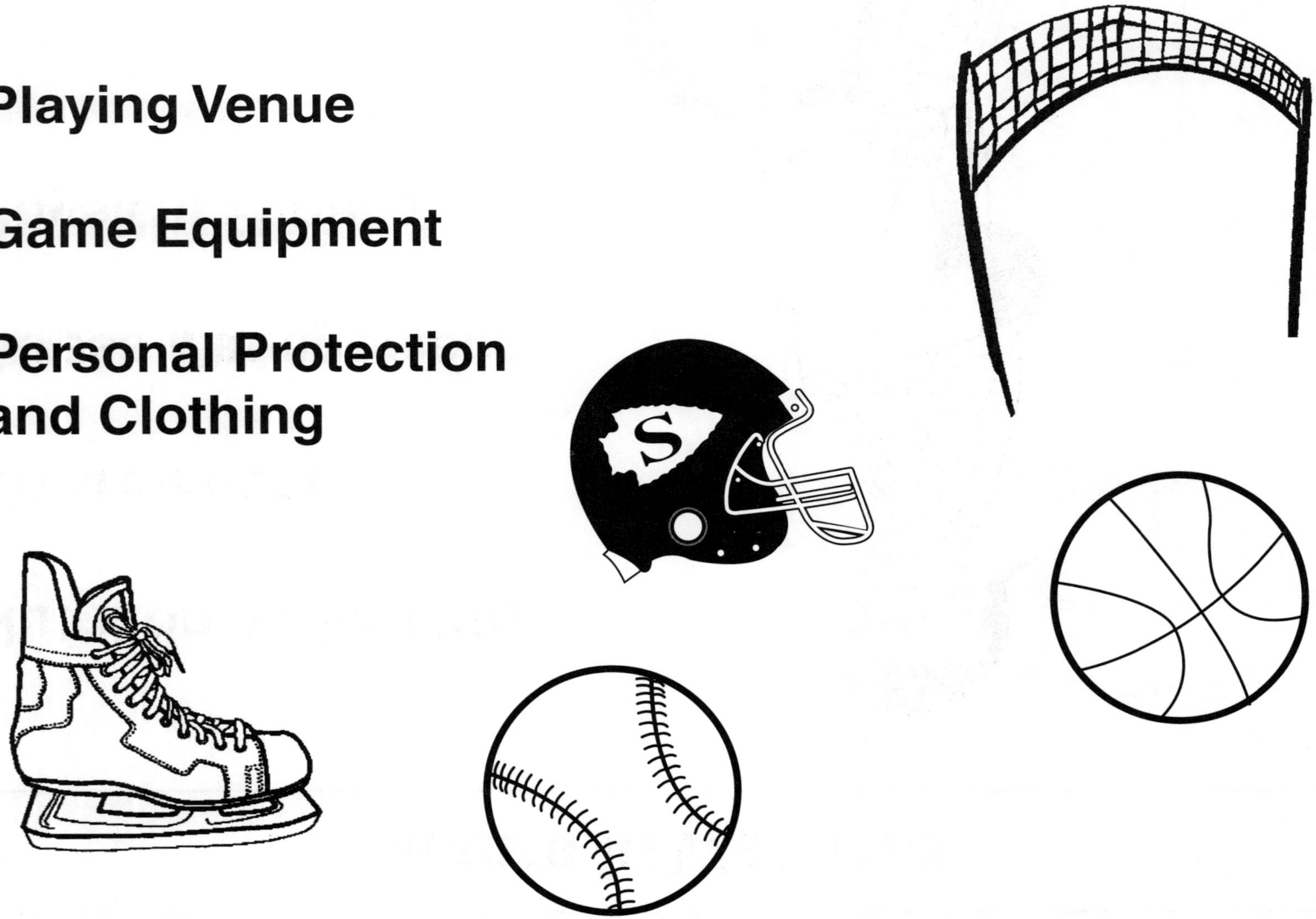

Goals of Medicine

Prevention

Diagnosis

Treatment

Section 10

Managing a Technological Enterprise

Teaching Materials

Text

Chapter 32, pp. 640–651
  Test Your Knowledge, pp. 650–651
  Applying Your Knowledge, p. 651

Chapter 33, pp. 652–673
  Test Your Knowledge, p. 673
  Applying Your Knowledge, p. 673

Chapter 34, pp. 674–683
  Test Your Knowledge, p. 683
  Applying Your Knowledge, p. 683

Section 10 Activities
  Activity 10A—Forming the Company, pp. 684–685
  Activity 10B—Operating the Company, pp. 685–686

Student Activity Manual

Activity 10A—Forming the Company, p. 119
Activity 10B—Operating the Company, pp. 121–129
Activity 10C—Using and Assessing Activity, p. 131
Taking Your Technology Knowledge Home, pp. 133–134

Teacher's Resources

Chapter 32 Quiz, pp. 310–311
Chapter 33 Quiz, pp. 317–318
Chapter 34 Quiz, pp. 326
Reproducible Masters
  32-1: Management, p. 312
  32-2: Functions of Management, p. 313
  32-3: Levels of Authority and Responsibility, p. 314
  33-1: Research and Development, p. 319
  33-2: Production, p. 320
  33-3: Marketing, p. 321
  33-4: Industrial Relations, p. 322
  33-5: Financial Affairs, p. 323
  34-1: Using Technological Products and Services, p. 327
  34-2: Assessing the Impacts of Technology, p. 328
Transparencies (Binder/CD only)
  33-1: Manufacturing Systems

Organization of Section 10

Up to this point in the course, the students have been studying technological practices that are used to design and use agricultural, construction, energy conversion, information and communication, manufacturing, medical, and transportation systems. These practices can be applied to all types of activities. Individuals can use them in the home workshop, and service clubs can use them to meet their goals. Most technology, however, is developed by one societal institution: industry. This institution is created and organized to provide economic value in growing and processing crops, communicating information, converting and applying energy, treating illnesses, transporting people and cargo, and producing products and structures.

In this section, the students will explore the concepts associated with organizing and operating an enterprise. They will address a number of topics, including:

- What is an enterprise?
- What is entrepreneurship?
- How are enterprises organized and financed?
- What is the role of management in an enterprise?
- What are the managed areas of activity in an enterprise?
- How do enterprises develop, produce, and market their products and services?
- How should people use and assess technology?

As the students address these concepts, they will study management and learn that it involves planning, organizing, actuating, and controlling activities, Figure 10-1. They will also learn that owners/managers organize a company by selecting the type of ownership, establishing the enterprise, and securing financing.

Next, the students will learn that society has five major institutions:

- Family: The basic unit of society that provides the foundation for social and economic actions

Functions of Management	
Planning	...setting goals and a course of action.
Organizing	...structuring the tasks and assigning work to people.
Actuating	...starting work and supervising employees.
Directing	...comparing results to the plan.

Figure 10-1

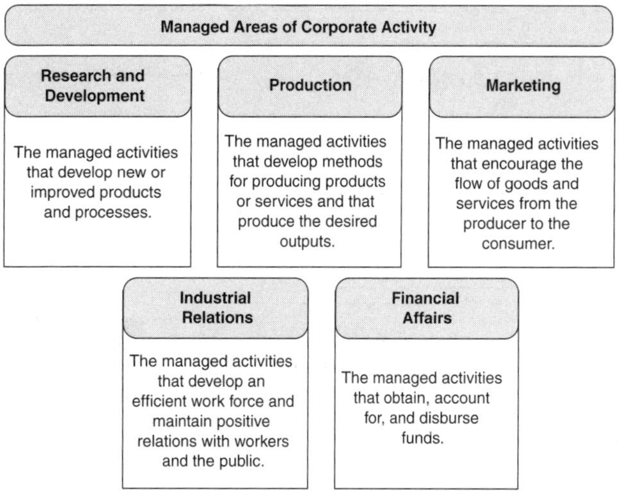

Figure 10-2

- Religion: The institution that develops and communicates values and beliefs about life and appropriate living
- Education: The institution that communicates information, ideas, and skills from one person to another and from one generation to another
- Political/Legal: The institution that establishes and enforces the rules of behavior and conduct for the society
- Economic: The institution that designs, produces, and delivers the basic goods and services required by society

Students then discover that industry is part of the economic institution and that industrial enterprises engage in five important managed activities, Figure 10-2. They are:

- Research and development: The managed activities that develop new or improved products and processes
- Production: The managed activities that develop methods for producing products or services and that produce the desired outputs
- Marketing: The managed activities that encourage the flow of goods and services from the producer to the consumer
- Industrial relations: The managed activities that develop an efficient work force and maintain positive relations with the workers and the public
- Financial affairs: The managed activities that obtain, account for, and disburse funds

As the students develop a conceptual understanding of industry, they organize model enterprises that let them apply their newly acquired knowledge. Three chapters and a multistage laboratory activity deliver this educational experience, Figure 10-3. The chapters are:

- Chapter 32—Organizing a Technological Enterprise
- Chapter 33—Operating Technological Enterprises
- Chapter 34—Using and Assessing Technology

The activities in this section consist of three phases for the enterprise and one using-and-assessing activity. The first phase is included in Activity 10A. Here, students are to organize an enterprise that will design, produce, and market a personalized calendar. They will sell space on any given date to promote a social or sporting event, a person's birthday, or other similar entries.

Activity 10B includes the remaining two phases, Figure 10-4. First, the students develop the financial plans, develop the product format, and promote and sell calendar entries. Then,

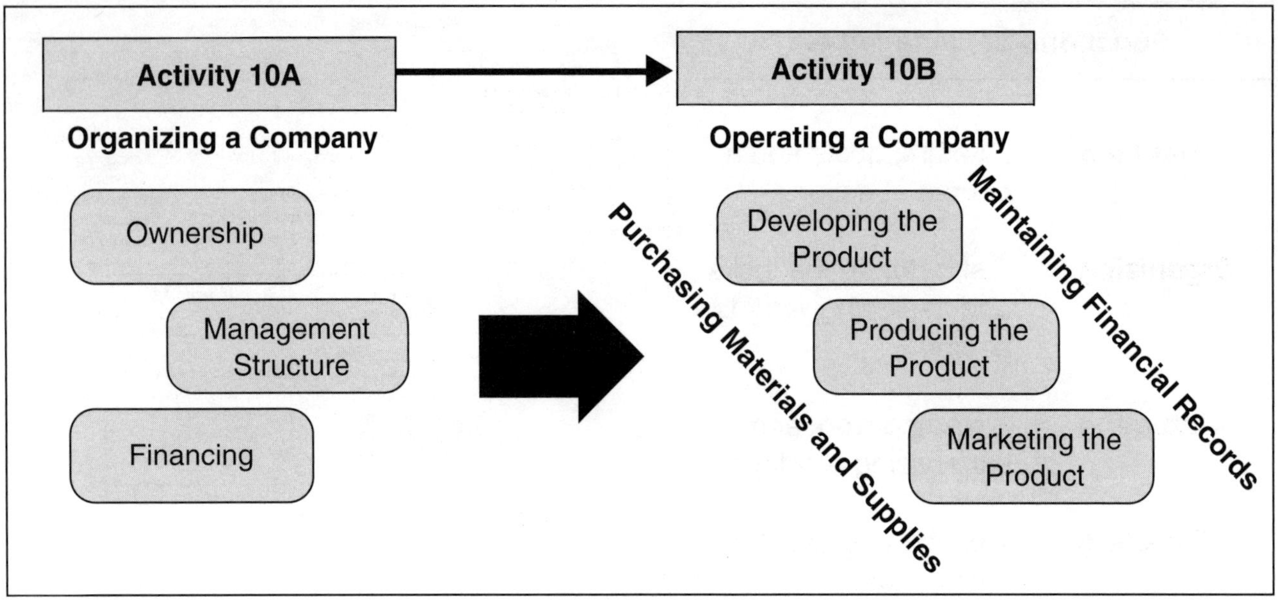

Figure 10-3

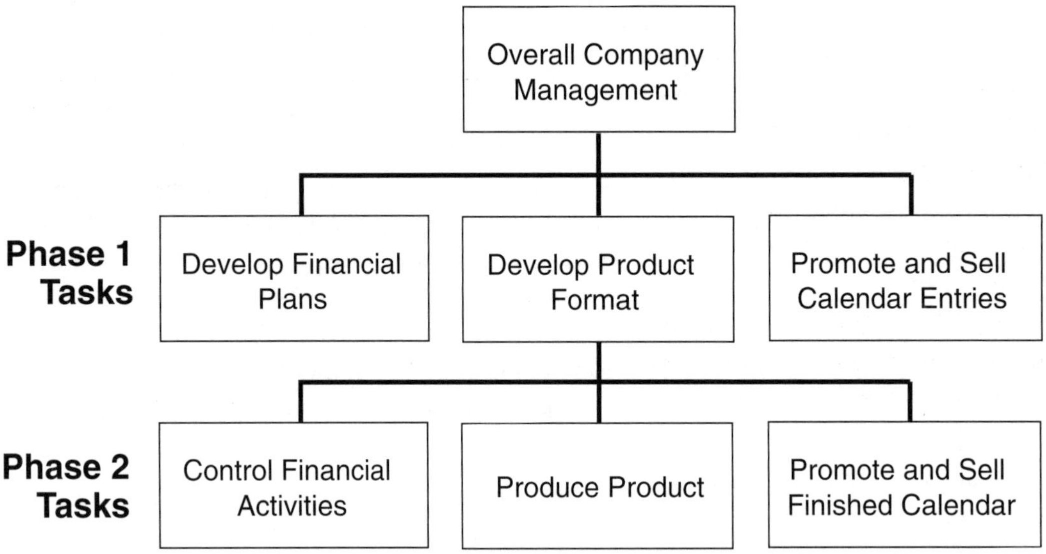

Figure 10-4

they control financial activities, produce the calendar, and promote and sell the product.

The using-and-assessing activity is contained in this guide. It allows the students to study how to use and assess properly the technological product they developed.

Presenting Section 10

Estimated length—4 weeks

Section 10 is designed to guide the students through the typical phases a new enterprise moves through. First, they will organize a company. Then, they will determine their financial needs and design a product. After that, they will produce and sell the product. Finally, they will determine how to use and assess the product. These activities help the students understand an enterprise as:

- A human-designed activity
- The primary source of new technology
- A dynamic organization in which many tasks are addressed simultaneously

- An organization that includes design, production, and marketing activities
- A major employer of managerial and technological talents
- A source of products people use and assess

The content of Section 10 is fairly general. Complete courses in organizing and operating an enterprise are often offered. Therefore, this four-week study, by its very nature, can only highlight selected enterprise activities. These activities include designing, producing, and selling products; purchasing materials to support production activities; maintaining financial records; and using and assessing products.

This content is supported by extensive laboratory activities that allow various students to perform different tasks supporting the overall mission of the enterprise: producing and selling a profitable product.

The twenty-day schedule, shown in Figure 10-5, has been developed for this section. During this period of time, you will notice that only three major topics are introduced through presentations and discussions. The majority of time is devoted to the students operating their enterprise.

To prepare to present this section, you should:

1. Read Chapters 32, 33, and 34 in *Technology.*
2. Prepare discussions on:
   a. Organizing a technological enterprise (Chapter 32).
   b. Financing enterprises (Chapter 32).
   c. Operating technological enterprises (Chapter 33).
   d. Using and assessing products (Chapter 34).

   Note: This presentation can be divided into three short presentations: (1) research and development activities, (2) production activities, and (3) marketing activities.
3. Practice the selected laboratory activity.
4. Prepare teaching materials (transparencies or bulletin boards, for example) to support the content.
5. Determine which study questions and quiz questions will be used.

Section 10: Managing Technological Systems	
Day	Activity
1	Discussion: "Organizing a Technological Enterprise" (Chapter 32) Introduce laboratory activity 10A
2–3	Discussion: "Financing Enterprises" (Chapter 32) Complete laboratory activity 10A
4	Discussion: "Operating Technological Enterprises" (Chapter 33) Introduce laboratory activity 10B
5–18	Students operate their enterprise Demonstrate as needed: Calendar software Product duplication (printing) Producing advertisements
19	Discussion: "Using and Assessing Technology" Complete laboratory activity 10B
20	Summarize section

Figure 10-5

Presenting the Activities

Consider the following recommendations in presenting the activities you select for this section.

Activity 10A—Forming the Company

This activity should be student centered. Students should be the source of most of the ideas for the company and the judges of the ideas that will be implemented. To start the activity, present the students with the situation and challenge for the simulation. The one presented in the textbook suggests the following:

Students have mentioned that they need a way to be informed about important sporting and social events at school. They also want a way to publicize happenings that they think are important. From these comments, it has been decided to organize a company to produce a nine-month calendar that spans the school year and has selected days personalized.

To implement this activity you may want to:

1. Divide the class into groups of three to four students.
2. Have each group use a *Student Activity Manual* sheet 10A to:
   a. List the tasks it thinks will be needed to design, produce, and sell a personalized calendar.
   b. Develop an organizational chart that shows which levels of management will be responsible for seeing that these tasks are completed.
3. Have each group present its ideas while you or a student keep a record on the chalkboard or flip charts.
4. Have the class select the best organizational chart.

Activity 10B—Operating the Company

Before introducing this activity, study Figure 10-6. It shows the important activities that must be completed in order to move the calendar from a need to a profitable product.

This activity includes all the tasks needed to complete the design, production, and marketing of the calendar. It consists of three major phases. Phase 1 includes designing the calendar. Phase 2 includes the actions needed to develop the financial plans, sell the entries, and develop the production format. Phase 3 includes purchasing materials, producing the calendars, selling the calendars, and maintaining appropriate records. To organize the overall operation, the class should have a president, production manager, marketing manager, and finance manager (controller). These officers will oversee the company through its life.

During Phase 1, first demonstrate the calendar production software and have a number of different formats produced by the software available for the students to analyze. Then have the students discuss and decide on the format characteristics they want to use. During this phase of the activity, page one of *Student Activity Manual* sheet 10B should be completed. This page will be the design criteria for all other actions in the company.

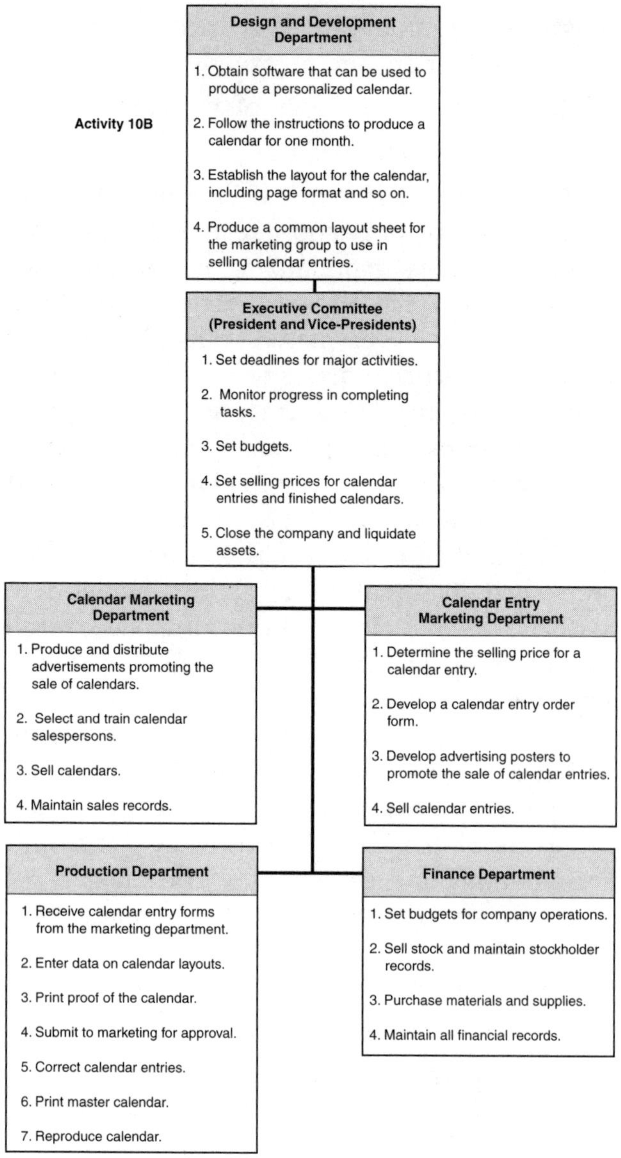

Figure 10-6

For Phase 2, assign teams of students to the production manager, marketing manager, and financial manager. Have the marketing team complete the calendar entry tasks on pages two and three of *Student Activity Manual* sheet 10B. Have the production team format the software and enter data as the marketing team sells entries. Have the finance team prepare a budget for the company and sell stock to finance operations, using pages four and five of *Student Activity Manual* sheet 10B.

For Phase 3, the production team should produce the calendar using an appropriate printing method. The marketing team should sell the calendar and maintain sales records, using pages six and seven of *Student Activity Manual* sheet 10B. The finance team should purchase material to produce the calendar and maintain the financial records, using pages eight and nine of *Student Activity Manual* sheet 10B. At the end of the activity, the company should be closed, financial reports should be prepared, and the company assets should be liquidated.

Activity 10C—Using and Assessing Activity

During this activity, the students will work in groups to determine how the product should be used and assess any impacts it could have on people, society, and the environment. They will complete this activity by following the directions on the appropriate sheet in the *Student Activity Manual.*

Alternate Activity—Section 10

This activity will have the students use a managed production system to produce a kite that can be sold to elementary and middle school students. Before the activity is introduced, collect a number of commercial kites and kite plans. Present the activity's three phases, Figure 10-7. Then the class should complete these phases.

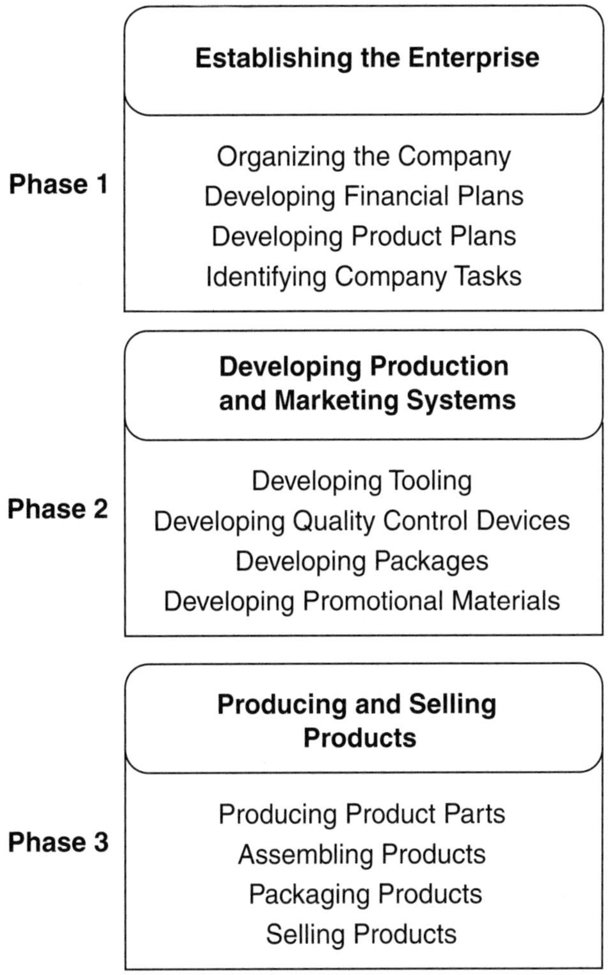

Figure 10-7

Phase 1—Establishing the Enterprise

During this phase, have the students work in groups to:

1. Organize the company.
2. Develop financial plans (budgets, break-even charts).
3. Develop product drawings (dimensioned sketches) by analyzing several existing kites or kite designs.
4. Identify the major tasks that will need to be completed to produce the kite. **Note:** As you review Phases 2 and 3, these tasks will become apparent.

Phase 2—Developing Production and Marketing Systems

During this phase, the class is divided into five groups as shown in the organizational chart in Figure 10-8. Each group is assigned one of the following tasks:

1. Develop a system to produce the sticks and paper cover for the kite (flowcharts or tooling, for example).
2. Develop a system to control the quality of the product (inspection gauges or quality control posters, for example).
3. Develop a product package (container or graphics, for example).
4. Develop product promotion (advertising) materials (flyers, radio spots for the school's public address system, or video commercials, for example).

Phase 3—Producing and Selling Products

During this phase of the activity, the students will complete the following tasks:

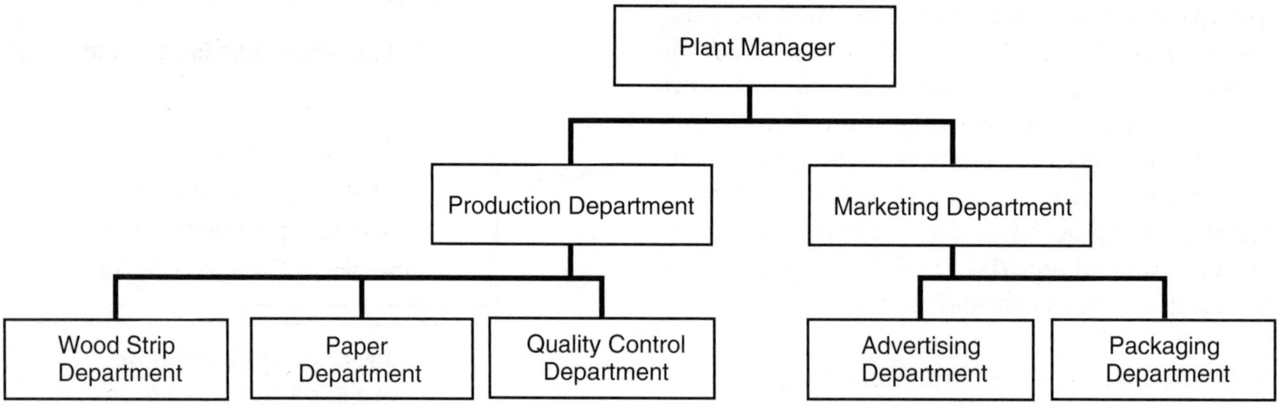

Figure 10-8

1. Produce kite parts.
2. Assemble the kits.
3. Package the kits.
4. Sell the kits to elementary and middle school students.

Reinforcing and Enriching

The following activities can be used to supplement and enrich the laboratory activities at the end of each textbook section and the Applying Your Knowledge activities included at the end of each textbook chapter.

1. A package could be developed to increase the product's marketing appeal.
2. A group of students could use their communication technology knowledge to develop a video or radio commercial promoting the product.

Taking Your Technology Knowledge Home

This take-home activity allows students to apply their knowledge of management systems to organize a social or civic event. Challenge the students to:

1. Work with their parents, guardians, neighbors, or youth group leader on an event for which they can prepare an organization and management system.
2. Determine the management functions that must be organized, such as production (day of the event management), marketing, financial affairs, and industrial relations (personnel assignment and training).
3. Prepare a management plan.
4. Present the plan to the event organizers.

Adjustments for Exceptional Students

The laboratory activities in this section, by their very nature, accommodate a wide range of student abilities and interests. The various activities include needs for artistic, mechanical, design, engineering, and managerial skills. Also, some of the tasks are routine, whereas others require more self-direction. Adjusting for individual differences requires that you carefully assign the tasks in terms of student abilities. This purposeful grouping of students will allow each student to bring his or her own special abilities to the group.

Evaluation Procedures

The understanding students have of technology and technological systems should be measured in several ways, including:

- Observation and critical listening during class presentations and discussions
- Worksheets and quizzes covering the cognitive material presented
- Open-ended questions about the learned material and how technology is developed
- Assessment of individual participation and contributions to group work
- Assessment of the quality and creativeness of the solutions to the assigned design/fabrication activities

Chapter 32

Organizing a Technological Enterprise

Objectives

After studying Chapter 32, the students will be able to:

- Explain how entrepreneurship and intrapreneurship differ.
- Define management.
- Identify the four functions of management.
- Name the three main forms of business ownership.
- List problems that can occur in the partnership form of business ownership.
- Explain why a corporation is said to be "like a person" in a legal sense.
- Define the role of a board of directors in a corporation.
- Describe the differences between equity financing and debt financing.

Answers to Test Your Knowledge *Questions, Text Pages 650–651*

1. Intrapreneurship
2. b. Authority
3. Actuating is initiating the work related to the action plan.
4. An educational television station would be considered a public enterprise if it is not organized and operated to create a profit.
5. Proprietorship, partnership, and the corporation
6. The corporation
7. d. File articles of incorporation
   b. Receive corporate charter
   a. Prepare bylaws
   c. Elect corporate directors
8. True
9. d. All of the above
10. Stock, bonds

Answers to the Chapter Quiz, Teacher's Resource

1. True
2. True
3. False
4. False
5. True
6. True
7. False
8. True
9. False
10. a. Stockholders
11. c. Sole proprietorship
12. c. Selling bonds
13. c. Stockholders
14. b. Bylaws
15. The main difference is that companies do not use a one-person/one-vote rule. Instead, each stockholder has as many votes as the shares of stock they own.

Captions for Reproducible Masters

Reproducible Master 32-1. Management—Management is the act of planning, directing, and evaluating any activity.

Reproducible Master 32-2. Functions of Management—Listed are the four important functions performed by management.

Reproducible Master 32-3. Levels of Authority and Responsibility—Good communication is important between these various levels of management.

Organizing a Technological Enterprise

Name ______________________________ Period ______________

Date ______________________________ Score ______________

Chapter 32 Quiz

True/False: If a statement is true, write *True* in the blank. If a statement is false, write *False* in the blank.

_________ 1. One does not need to start a new business to be an entrepreneur.

_________ 2. One of the advantages of a corporation is that each owner cannot lose more money than he or she invested in the company.

_________ 3. Congress establishes the rules for forming a corporation.

_________ 4. A corporate charter issued in one state might not be honored by another state.

_________ 5. An owner of a proprietorship has complete control over the business.

_________ 6. People with an entrepreneurial spirit who work within an existing company are called intrapreneurs.

_________ 7. In a corporation, the owners have ultimate control.

_________ 8. Selling shares of stock in a company is called equity financing.

_________ 9. Owners of a corporation are called bondholders.

Multiple Choice: Choose the best answer and write the corresponding letter in the blank.

_____ 10. Publicly held enterprises are owned by:
- a. stockholders.
- b. government agencies.
- c. two or more individuals.
- d. None of the above.

_____ 11. The easiest type of company to form is a:
- a. partnership.
- b. corporation.
- c. sole proprietorship.
- d. public enterprise.

_____ 12. Debt financing involves raising money by:
- a. converting a privately held enterprise into a publicly held enterprise.
- b. selling shares of stock.
- c. selling bonds.
- d. Both a. and b.

_____ 13. Dividends are profits paid to:
- a. bondholders.
- b. employees.
- c. stockholders.
- d. the board of directors.

(Continued)

Name ____ Period ____________________

_____ 14. The general rules by which a company will operate are called its:
a. corporate charter.
b. bylaws.
c. articles of incorporation.
d. All of the above.

Short Answer: In the space provided, write a brief answer to the following question.

15. Briefly describe how the election of a board of directors differs from the elections used to elect a political candidate.

____

____

____

____

____

Management

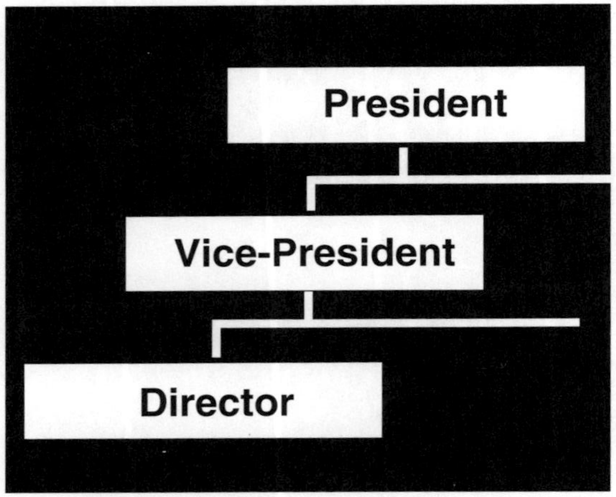

Guiding and directing company activities so they operate in an efficient manner.

Levels of Authority and Responsibility

Function

Areas of Managed Activity

Functions of Management

Plan

Setting Goals and Courses of Action

Organize

Structuring and Staffing the Organization

Actuate

Assigning Tasks and Supervising Work

Control

Measuring Results against Plans

Levels of Authority and Responsibility

Top Management

↓ ↑

Middle Management

↓ ↑

Operating Management

Chapter 33
Operating Technological Enterprises

Objectives

After studying Chapter 33, the students will be able to:

- Define an economic enterprise.
- Diagram the relationships among the five main managed areas of activity within a technological enterprise.
- Explain the difference between basic research and applied research.
- Name the four common systems of manufacturing.
- Define the term "quality control."
- Describe the channels of distribution used to move products from producer to the consumer.
- List the major types of programs that are included in industrial relations.
- Describe the different types of employee training.
- Explain how expenses, income, and profit are related.
- Identify the steps in the industry-consumer product cycle.

Answers to Test Your Knowledge *Questions, Text Page 673*

1. Industry is the area of economic activity that uses resources and systems to produce products, structures, and services with intent to make a profit.
2. e. Financial affairs
3. b. Production
4. a. Research and development
5. c. Marketing
6. b. Production
7. d. Industrial relations
8. e. Financial affairs
9. Applied
10. Custom manufacturing, intermittent (job-lot) manufacturing, continuous manufacturing, and flexible manufacturing
11. Materials and purchased parts, end product
12. False
13. True
14. Income
15. Consumers select, use, maintain, and discard the products.

Answers to the Chapter Quiz, Teacher's Resource

1. False
2. True
3. True
4. True
5. False
6. True
7. False
8. c. Marketing activities
9. b. Product development
10. a. Intermittent manufacturing
11. b. Apprenticeship training
12. c. Retained earnings
13. d. Both a. and c.
14. b. Family, religion, education, political/legal, and economic
15. *Research and development:* The activities that might result in new or improved products and processes. *Production:* The activities that develop methods for producing products or services and the activities that produce the desired outputs. *Marketing:* The activities that encourage the flow of goods and services from the producer to the consumer. *Industrial relations:* The activities that develop an efficient work force and the activities that maintain positive relations with the workers and the public. *Financial affairs:* The activities that obtain, account for, and disburse funds.

Captions for Reproducible Masters

Reproducible Master 33-1. Research and Development—Research and development convert what the human mind envisions into physical products and services.

Reproducible Master 33-2. Production—Production manufactures or constructs the physical item.

Reproducible Master 33-3. Marketing—Marketing efforts promote, sell, and distribute products, structures, and services.

Reproducible Master 33-4. Industrial Relations—Shown are the three important areas of industrial relations: public relations, employee relations, and union (labor) relations.

Reproducible Master 33-5. Financial Affairs—Financial affairs direct and control a company's monetary operations.

Caption for Transparency Master (Binder/CD Only)

Transparency Master 33-1. Manufacturing Systems—Shown are five important areas of managed activities that change ideas into products, structures, and services. Each area relates with the others.

Operating Technological Enterprises

Name ______________________________ Period ______________

Date ______________________________ Score ______________

Chapter 33 Quiz

True/False: If a statement is true, write *True* in the blank. If a statement is false, write *False* in the blank.

_________ 1. Basic research finds a use for knowledge developed through applied research.

_________ 2. Continuous manufacturing has the advantage of low cost but is fairly inflexible.

_________ 3. When random inspection is used, a sample is selected for inspection. If the sample does not meet quality standards, the entire run is rejected.

_________ 4. Sales planning develops selling methods and selects and trains sales personnel for their efforts.

_________ 5. Labor relations include programs to recruit, develop, and reward a company's employees.

_________ 6. A company's public relations department enhances the company's image and its ability to work with the community and government agencies.

_________ 7. All economic activities are aimed at developing technology.

Multiple Choice: Choose the best answer and write the corresponding letter in the blank.

_____ 8. Managed activities that encourage the flow of goods from producers to consumers are called:
 a. production activities.
 b. industrial relations.
 c. marketing activities.
 d. research and development.

_____ 9. Which of the following applies knowledge to design new products?
 a. Product and structure engineering
 b. Product development
 c. Process development
 d. None of the above

_____ 10. In which type of manufacturing process do all parts move through the process in a single batch?
 a. Intermittent manufacturing
 b. Flexible manufacturing
 c. Continuous manufacturing
 d. Custom manufacturing

_____ 11. Highly skilled workers are generally trained through what process?
 a. On-the-job training
 b. Apprenticeship training
 c. Classroom training
 d. None of the above

(Continued)

Name ______________________________

_____ 12. Profits reinvested in a company to expand its operations are termed:
 a. income.
 b. dividends.
 c. retained earnings.
 d. None of the above.

_____ 13. Which of the following categories of activities are product centered?
 a. Marketing
 b. Industrial relations
 c. Research and development
 d. Both a. and c.

Short Answer: In the space provided, write a brief answer to each of the following questions.

14. List the five institutions found in nearly all societies.

____

____

____

____

____

15. List and define the five managed areas of activity found in most technological enterprises.

____

____

____

____

____

Research and Development

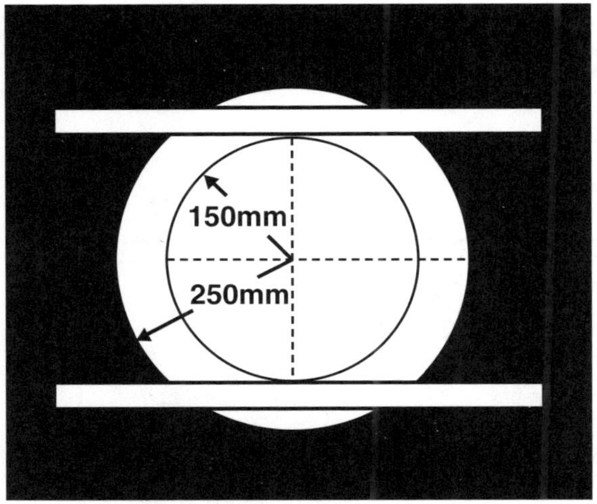

Develops and specifies new and improved products and processes.

Research

Development

Engineering

Production

Develops systems to produce products to stated standards.

Manufacturing Engineering

Manufacturing

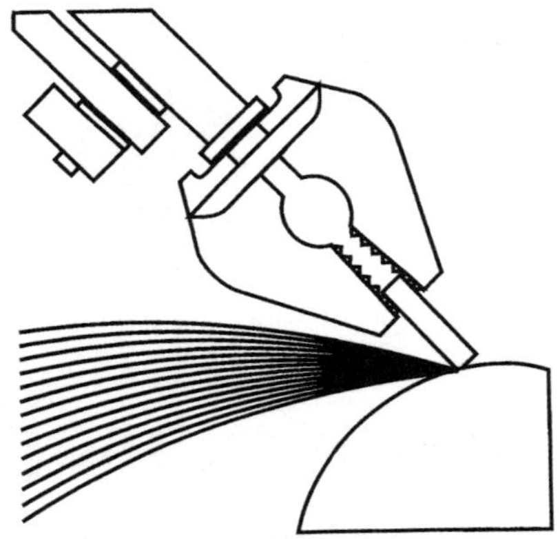

Quality Control

Production Planning and Control

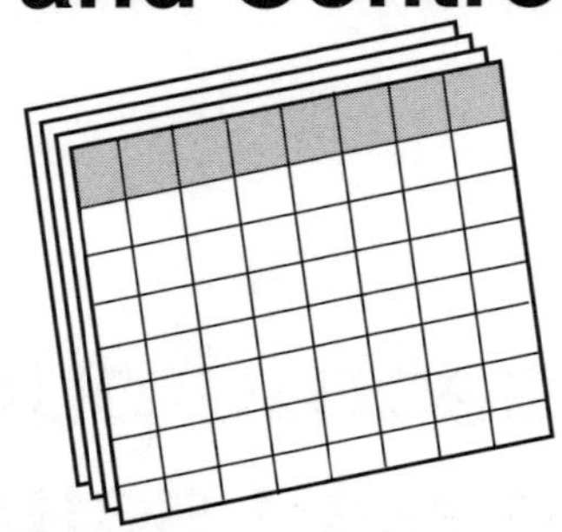

Marketing

Encourages the flow of goods from the producer to the consumer.

Market Research

Advertising

Sales

Distribution

Industrial Relations

Ensures that the relationships among management, the workers, and the public are positive.

Public Relations

Employees Relations

Union Relations

Financial Affairs

Directs and controls the company's monetary operations.

Finance

Control

Purchasing

Chapter 34
Using and Assessing Technology

Objectives

After studying Chapter 34, the students will be able to:

- Describe how using a technological product is different from using a technological service.
- Explain how people can effectively select an appropriate product or technological service.
- Indicate where people can obtain information about operating products.
- Explain how products should be properly discarded.
- Describe ways societies assess technological advancements.

Answers to Test Your Knowledge *Questions, Text Page 683*

1. Selecting appropriate devices and systems, operating devices and systems properly, maintaining devices and systems, repairing broken parts and systems, and disposing of worn out or obsolete devices and systems
2. Need
3. Student answers should describe the select-use-evaluate model.
4. Assessment
5. First, the underlying goal for the technology needs to be identified. Second, a way to measure the success in meeting the goal needs to be developed. Third, the assessment needs to monitor and measure the impact of the technological innovation. Fourth, people must review the data and draw conclusions. Finally, the assessment group must prepare a report of its findings.

Answers to the Chapter Quiz, Teacher's Resource

1. True
2. False
3. True
4. True
5. True
6. False
7. True
8. a. Determine the goal of the technology
9. d. All of the above
10. Technological products are the artifacts that people build. Technological services are outputs that we use but do not own.

Captions for Reproducible Masters

Reproducible Master 34-1. Using Technological Products and Services—Shows the steps in using both a product and a service.

Reproducible Master 34-2. Assessing the Impacts of Technology—Shows a five-step process to assess technological impacts.

Using and Assessing Technology

Name ____ Period ____________________

Date ____ Score ____________________

Chapter 34 Quiz

True/False: If a statement is true, write *True* in the blank. If a statement is false, write *False* in the blank.

_________ 1. Technological services are outputs that customers use but do not own.

_________ 2. Selecting and using a technological service require more actions by the customer.

_________ 3. Repairing devices includes making major operating adjustments.

_________ 4. Before operating an unfamiliar device, you should determine how to care for and maintain it.

_________ 5. If a product cannot be reused or recycled, it should be sent to a landfill.

_________ 6. The report created by the assessment group should state facts but should not suggest an action to be taken.

_________ 7. Some products need little or no training to be able to use properly.

Multiple Choice: Choose the best answer and write the corresponding letter in the blank.

_____ 8. Which of the following is a necessary, fundamental step in assessing technology?
- a. Determine the goal of the technology
- b. Determine if the technology is a logical progression of an existing technology
- c. Survey a random sample of people to determine what they like or don't like about the technology
- d. None of the above

_____ 9. Which of the following is an acceptable method for learning the proper operation of an unfamiliar device?
- a. Attending a class or seminar
- b. Asking a person skilled in using the device
- c. Reading the owner's manual
- d. All of the above

Short Answer: In the space provided, write a brief answer to the following question.

10. Explain the difference between technological products and technological services.

____

____

____

____

____

Using Technological Products and Services

Technological Device

Select → Operate → Service → Repair → Dispose

Technological Service

Select → Use → Select

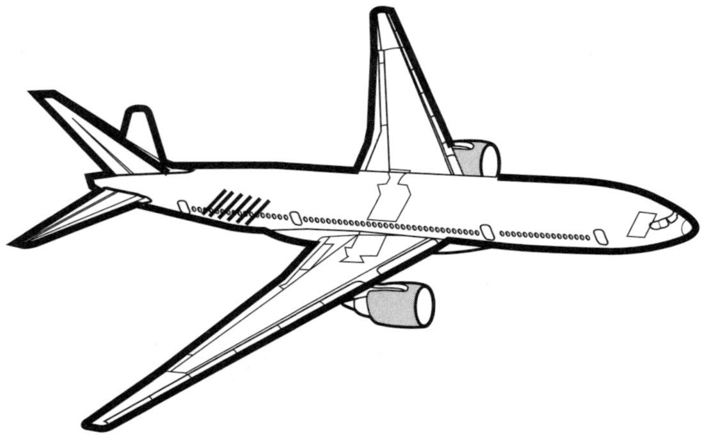

Assessing the Impacts of Technology

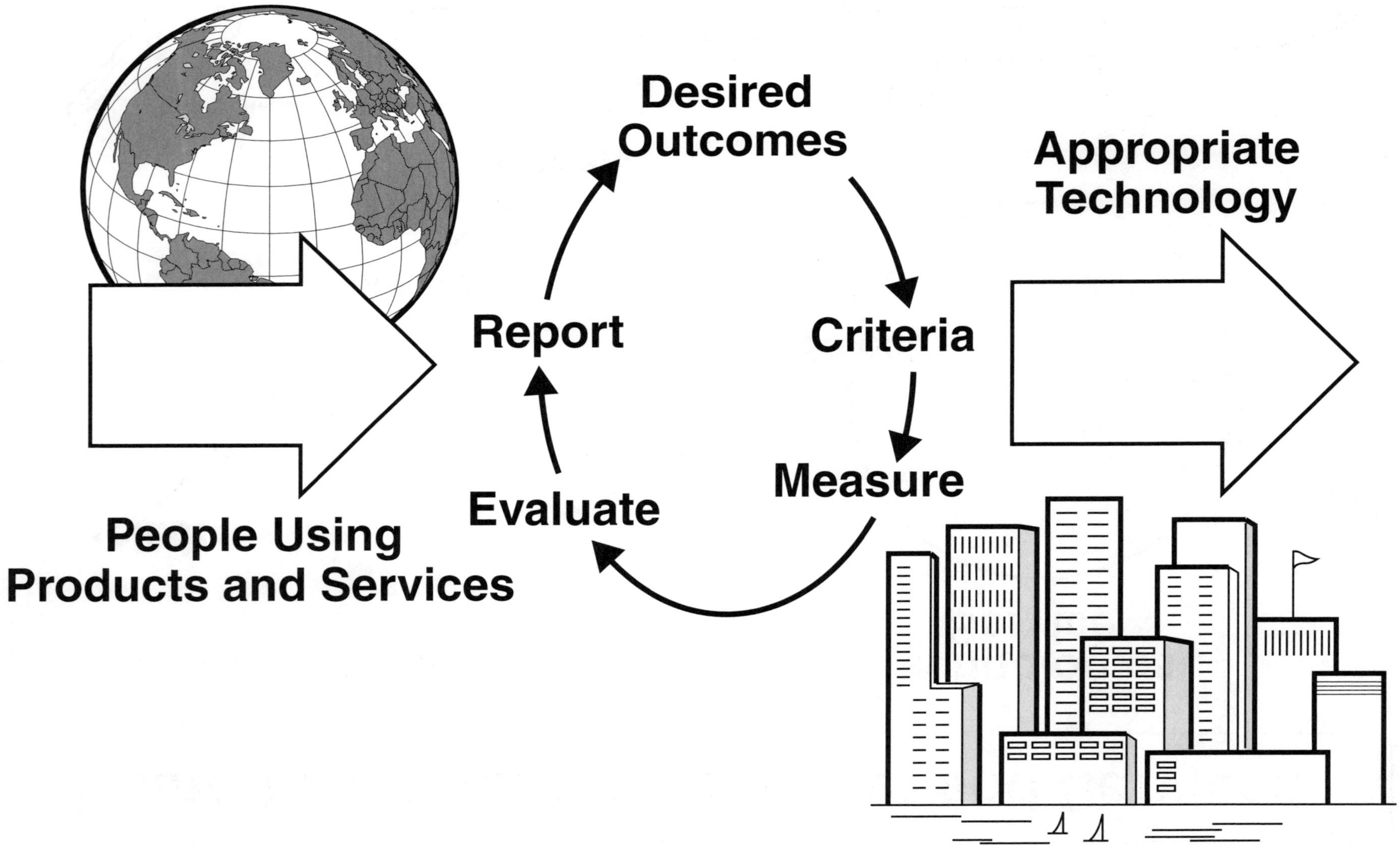

Section 11

Technological Systems in Modern Society

Teaching Materials

Text

Chapter 35, pp. 690–704
- Test Your Knowledge, pp. 703–704
- Applying Your Knowledge, p. 704

Chapter 36, pp. 706–721
- Test Your Knowledge, pp. 720–721
- Applying Your Knowledge, p. 721

Section 11 Activities
- Activity 11A—Design Problem, p. 722
- Activity 11B—Production Problem, pp. 722–723

Student Activity Manual

Activity 11A—Design Problem, pp. 137–138
Activity 11B—Production Problem, pp. 139–140

Teacher's Resources

Chapter 35 Quiz, pp. 334–335
Chapter 36 Quiz, p. 340
Reproducible Masters
- 35-1: Features of Futuring, p. 336
- 35-2: Some Challenges to the Environment, p. 337
- 36-1: Some Technology-Based Jobs, p. 341
- 36-2: Job Requirements, p. 342

Transparencies (Binder/CD only)
- 35-1: Switching Resources
- 35-2: Pollution

Organization of Section 11

This last section studies technology and its design, operation, and impacts. It is not, however, to be considered less important than the other sections. In fact, in the final analysis, it might be the most important because it focuses on technology as it impacts people, society, and the environment.

It involves a dual view of technology. The first view is an outward, encompassing view of technology. It looks at technology and society. It also explores its impacts on people as a group and their environment.

The second view is an inwardly looking personal view, dealing with technology and the individual. It explores how technology makes life different for each person. It also introduces the student to the demands and hierarchy of technological careers.

During Section 11, the students will learn about futuring, which is a process that helps people choose wisely among alternate, and sometimes conflicting, courses of action. They will learn that futuring has five distinct features, Figure 11-1:

- Alternate avenues—possible answers rather than the answer
- Different futures—seeking an entirely new future
- Rational decision making—using logical thinking and considering the consequences to make decisions
- Designing the future—predicting a possible future that can be created
- Interrelationships—seeing alternatives, cross-impacts, and leaps forward

The students will use some of these ways of thinking to address common technological problems, such as energy conservation, pollution, and global warming.

Later in the section, the students will take personal looks at technology as they explore a wide range of careers, such as those of the production worker, technician, engineer, technologist, and manager. Then they will learn how to select a career that meets their abilities and desired lifestyles.

To complete the study, the students will complete two chapters and two laboratory activities. The chapters are:

- Chapter 35—Technology: A Societal View
- Chapter 36—Technology: A Personal View

Features of Futuring
Alternate Avenues Different Futures Rational Decision Making Designing the Future Interrelationships

Figure 11-1

As with most of the sections, two different types of activities are provided. If you want the students to address the content from a problem-solving approach, select Activity 11A—Design Problem. If you want a more structured approach, select Activity 11B—Production Problem.

Presenting Section 11

Estimated length—1 week

Section 11 is designed to help the students understand technology as a neutral force that can be made positive or negative when applied by people. For example, nuclear reactions are neutral. However, people can apply them to generate electricity, which is a positive force, or to create a nuclear winter through weapons, which is a negative force.

The content of Section 11 lends itself to a lecture/discussion strategy, where the students could simply discuss the impacts of technology on themselves and their environments. More meaningful learning, however, is accomplished by using realistic laboratory activities to reinforce the discussions.

The five-day schedule, shown in Figure 11-2, has been developed for this section. During this period of time, you will notice that two major focuses are used for the study—the individual and society/environment. To prepare to present this section, you should:

1. Read Chapters 35 and 36 in *Technology.*
2. Prepare discussions on:
   a. Technology, society, and the environment. (Chapter 35)
   b. You and technology. (Chapter 36)
3. Select a laboratory activity.
   a. Activity 11A—This is an open-ended activity where students select a technological issue, analyze it, and suggest an action plan to improve it.
   b. Activity 11B—This is a structured activity where students develop 60-second television commercials that promote a positive technological action.
4. Practice the selected laboratory activity.
5. Prepare teaching materials (transparencies or bulletin boards, for example) to support the content.
6. Determine which study questions and quiz questions will be used.

Section 11: Technological Systems in Modern Society	
Day	Activity
1–2	Discussion: "Technology, Society, and the Environment" (Chapter 35) Introduce laboratory activity
3	Students work on laboratory activity
4	Discussion: "You and Technology" (Chapter 36) Students work on laboratory activity
5	Students complete laboratory activity Summarize section and course

Figure 11-2

Presenting the Activities

Consider the following recommendations in presenting the activities you select for this section.

Activity 11A—Design Problem

To present this activity, first set the stage by having the students discuss technology as a neutral factor. Then encourage them to give examples of how applications of various technologies either benefit or harm people, society, and the environment.

Suggest that each person should strive to use technology wisely to make the future better, protect the environment, and help people live in harmony with nature. This suggestion should lead into the challenge, which is to identify an issue important to people today that is impacted by technology and develop a plan to maximize the positive features and reduce the negative impacts.

To facilitate this activity:

1. Divide the class into groups of three to four students.

2. Have each group:
   a. Identify three to five different technological issues, such as ozone depletion, hazardous waste, sanitary waste, smog, and acid rain.
   b. Select one issue to explore.
   c. Bring in at least three magazine or newspaper articles on the issue.
   d. List the positive and negative factors related to the issue.
   e. Develop a plan to maximize the positive features and reduce the negative impacts.
3. If time permits, have the groups develop posters, pamphlets, or displays of their findings.

Activity 11B—Production Problem

This structured activity encourages the students to explore the impacts of technology on individuals and society. The students are challenged to produce television commercials that will inform people about a technological issue or cause them to act differently as they use technology. The specific challenge outlined in the textbook suggests that the student is the communication director of a citizen's group. The group has determined that people are not participating in the local recycling program.

The student is to design a 60-second public-service television commercial. The commercial needs to explain the importance of recycling to the community. It needs to identify the items that can be recycled and the benefits for recycling them.

To complete this activity, the students will need *Student Activity Manual* sheet 11B (storyboard form); pencils or felt tip pens; and a video camera, recorder, and monitor.

To design and produce the commercial, the students should follow the procedure outlined in the textbook. If time is restricted, the students could produce a radio commercial, which requires less time and technical expertise.

Reinforcing and Enriching

The following activity can be used to supplement and enrich the laboratory activities at the end of each textbook section and the Applying Your Knowledge activities included at the end of each textbook chapter. Activity 11B—Production Problem could be reinforced by working with the social studies and language arts teachers so that the research and writing activities could be coordinated with those classes. Your class could handle the production activities.

Adjustments for Exceptional Students

Both laboratory activities are designed to be done rather quickly. The level of sophistication can be adjusted to accommodate students of all ability levels. Gifted students should have more creative answers and possibly produce a more complex solution. Lower achievers can still solve the problem but with solutions that fit within their ability level. Also, by grouping students of various interests and abilities, each student can bring his or her own special talents to the group. Some students will have more artistic abilities, others might have more dexterity, and still others might have more abstract reasoning ability.

Evaluation Procedures

The understanding that students have of technology and technological systems should be measured in several ways, including:

- Observation and critical listening during class presentations and discussions.
- Worksheets and quizzes covering the cognitive material presented.
- Open-ended questions about the learned material and how technology is developed.
- Assessment of individual participation and contributions to group work.
- Assessment of the quality and creativeness of the solutions to the assigned design/production activities.

Chapter 35
Technology: A Societal View

Objectives

After studying Chapter 35, the students will be able to:

- List some destructive natural forces that affect human lives.
- Give some examples of how technology is used to help to control natural forces.
- Describe the global impact of automobile use in North America.
- Distinguish between divergent thinking and convergent thinking as futuring tools.
- Describe the four major types of futures that are considered in futures research.
- List the three forces that are causing a condition called an environmental crisis.
- Trace the movement of the center of technical knowledge and power from east to west through human history.
- Identify several specific actions companies must take to remain competitive in a world economy.

Answers to Test Your Knowledge *Questions, Text Pages 703–704*

1. True
2. Evaluate individually.
3. Natural
4. The automobile
5. Divergent thinking explores all possible and impossible solutions. Convergent thinking focuses attention on a single solution.
6. b. Economic
7. Technology
8. Overpopulation, resource depletion, and pollution
9. Industrial
10. CAD (computer-aided design), CIM (computer-integrated manufacturing), and just-in-time (JIT) inventory control.

Answers to the Chapter Quiz, Teacher's Resource

1. False
2. True
3. False
4. True
5. True
6. c. Social future
7. b. Water
8. Technology is developed to help people modify and control the natural world.
9. Futuring or futures research
10. Evaluate individually.

Captions for Reproducible Masters

Reproducible Master 35-1. Features of Futuring—Futuring may be done using alternate avenues, different futures, rational decision making, designing the future, and interrelationships.

Reproducible Master 35-2. Some Challenges to the Environment—Some of the challenges that people place on the environment are pollution, resource depletion, and overpopulation.

Captions for Transparency Masters (Binder/CD Only)

Transparency Master 35-1. Switching Resources—Our limited supply of exhaustible resources will require us to rely more on renewable and inexhaustible resources in the future.

Transparency Master 35-2. Pollution—Pollution affects the land, water, and air that we depend on for life.

Technology: A Societal View

Name ______________________________ Period ____________

Date ______________________________ Score ____________

Chapter 35 Quiz

True/False: If a statement is true, write *True* in the blank. If a statement is false, write *False* in the blank.

_________ 1. A futurist views the future as a refinement of the present.

_________ 2. Traditional planners rely on statistical projection much more than do futurists.

_________ 3. Petroleum is a renewable resource that we are in no danger of depleting.

_________ 4. Technology has added to the problem of overpopulation by increasing human life expectancy.

_________ 5. Energy resources are not the only resources with a finite supply; some materials used in manufacturing are also exhaustible resources.

Multiple Choice: Choose the best answer and write the corresponding letter in the blank.

_____ 6. Which type of future focuses on people's relationships with each other?
  a. Human-psyche future
  b. Biological future
  c. Social future
  d. Technological future

_____ 7. Which of the following is an example of an inexhaustible energy source?
  a. Natural gas
  b. Water
  c. Wood
  d. Coal

Short Answer: In the space provided, write a brief answer to each of the following questions.

8. Why is technology developed?

____

____

____

____

____

(Continued)

Name ______________________________

9. What do we call the research techniques that try to predict the future?

____

10. List two major problems that could be addressed by futuring.

____

____

Features of Futuring

- Alternate Avenues
- Different Futures
- Rational Decision Making
- Designing the Future
- Interrelationships

Some Challenges to the Environment

Pollution

Resource Depletion

Overpopulation

Chapter 36
Technology: A Personal View

Objectives

After studying Chapter 36, the students will be able to:

- Describe how jobs have evolved from the colonial period, through the industrial revolution, to the information age.
- Discuss the typical levels of technology-based jobs and the types of education required for each.
- List the factors that should be considered in selecting a job.
- Identify the three factors that affect job satisfaction.
- Describe the ways that individuals exercise control over technology.
- List some technological activities introduced as fiction that have become actual processes.

Answers to Test Your Knowledge *Questions, Text Pages 720–721*

1. 90
2. Immigrants and out-of-work farmers
3. True
4. Production worker, technicians, technologists, engineers, and managers
5. Technologist, engineer, and manager
6. d. All of the above
7. People
8. False
9. Through the role of consumer, through political power, and through activism
10. Evaluate individually.

Answers to the Chapter Quiz, Teacher's Resource

1. True
2. False
3. False
4. True
5. False
6. True
7. False
8. True
9. Professional management, division of labor, continuous manufacturing techniques, material handling devices, and interchangeable parts
10. Evaluate individually.

Captions for Reproducible Masters

Reproducible Master 36-1. Some Technology-Based Jobs—Production workers, technicians, technologists, engineers, and managers need technological knowledge and skills.

Reproducible Master 36-2. Job Requirements—Jobs require skills in working with people, data, and machines.

Technology: A Personal View

Name ______________________________ Period ____________

Date ______________________________ Score ____________

Chapter 36 Quiz

True/False: If a statement is true, write *True* in the blank. If a statement is false, write *False* in the blank.

_________ 1. People put out of work by the mechanization of farming in the 1800s provided a large portion of the labor supply necessary for the Industrial Revolution.

_________ 2. Computer-controlled machines can replace highly skilled workers very easily and help create a flexible factory.

_________ 3. Production workers are more highly trained than technicians and often possess an associate degree.

_________ 4. Every job requires working with data, machines, or people.

_________ 5. Scientists and others who develop and produce technology are the only ones who have control over technology.

_________ 6. A good consumer should be able to select, operate, maintain, and recycle or dispose of technological devices properly.

_________ 7. Most people were their own bosses during the Industrial Revolution.

_________ 8. People should match their job with their belief system.

Short Answer: In the space provided, write a brief answer to each of the following questions.

9. List five characteristics of the factories that appeared during the Industrial Revolution.

10. Describe an important concern of yours that technology has caused. How could it be addressed?

Some Technology-Based Jobs

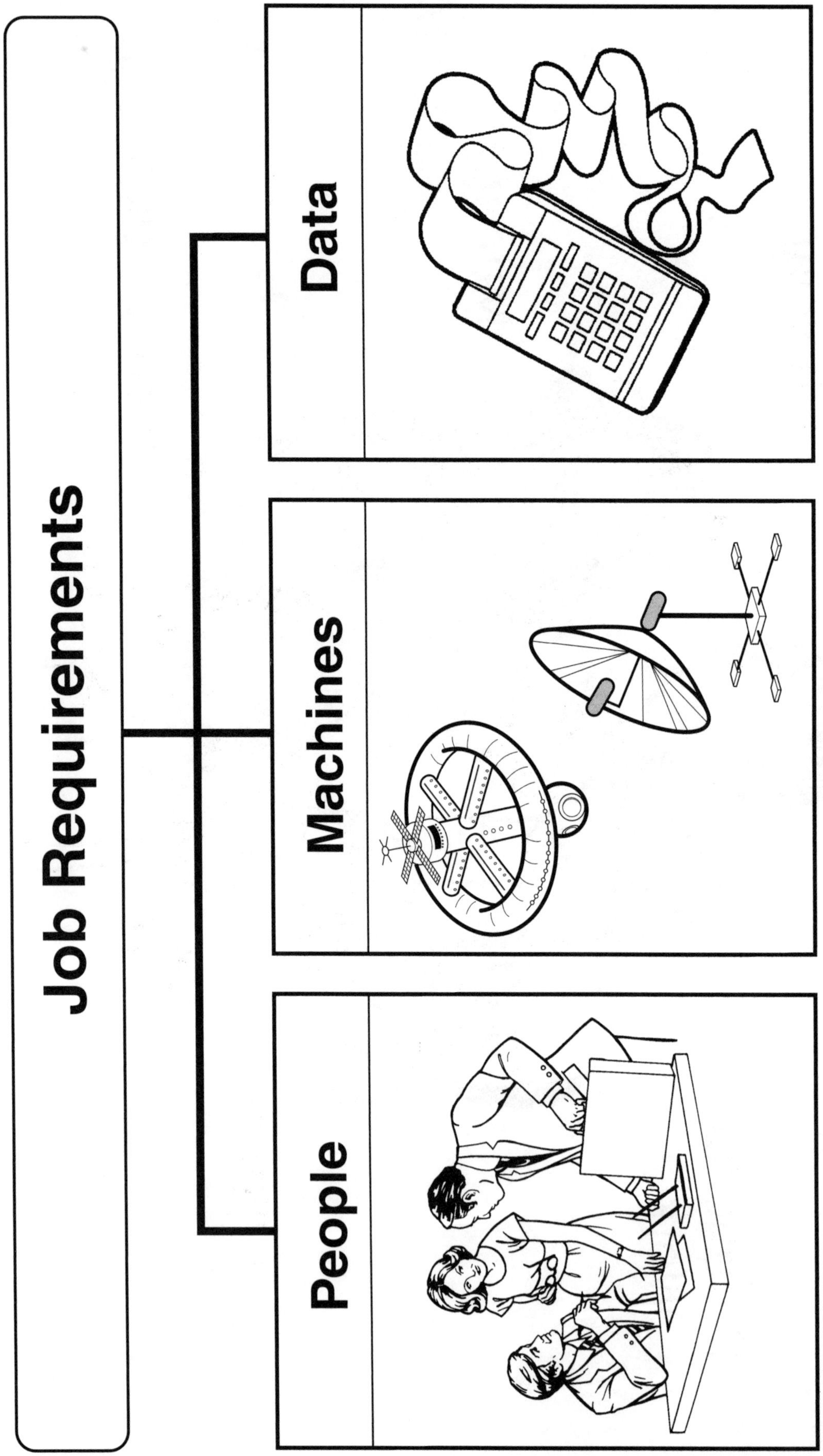
Job Requirements
People
Machines
Data